# Agrarianism as Modernity in 20th-Century Europe

# Agrarianism as Modernity in 20th-Century Europe

*The Golden Age of the Peasantry*

Alex Toshkov

BLOOMSBURY ACADEMIC
LONDON • NEW YORK • OXFORD • NEW DELHI • SYDNEY

BLOOMSBURY ACADEMIC
Bloomsbury Publishing Plc
50 Bedford Square, London, WC1B 3DP, UK
1385 Broadway, New York, NY 10018, USA
29 Earlsfort Terrace, Dublin 2, Ireland

BLOOMSBURY, BLOOMSBURY ACADEMIC and the Diana logo are trademarks of Bloomsbury Publishing Plc

First published in Great Britain 2019
This paperback edition published in 2021

Copyright © Alex Toshkov, 2019

Alex Toshkov has asserted his right under the Copyright, Designs and Patents Act, 1988, to be identified as Author of this work.

Cover design by Anna Toshkova
Cover image © The spectre of famine approaches the well fed Aleksandar Stamboliyski the prime minister of Bulgaria in 1919 during the widespread European famine that followed in the wake of the First World War. (© Michael Nicholson/Corbis/Getty Images)

All rights reserved. No part of this publication may be reproduced or transmitted in any form or by any means, electronic or mechanical, including photocopying, recording, or any information storage or retrieval system, without prior permission in writing from the publishers.

Bloomsbury Publishing Plc does not have any control over, or responsibility for, any third-party websites referred to or in this book. All internet addresses given in this book were correct at the time of going to press. The author and publisher regret any inconvenience caused if addresses have changed or sites have ceased to exist, but can accept no responsibility for any such changes.

A catalogue record for this book is available from the British Library.

A catalog record for this book is available from the Library of Congress.

ISBN: HB: 978-1-3500-9055-2
PB: 978-1-3502-1667-9
ePDF: 978-1-3500-9056-9
eBook: 978-1-3500-9057-6

Typeset by Deanta Global Publishing Services, Chennai, India

To find out more about our authors and books visit www.bloomsbury.com and sign up for our newsletters.

*To my mother, without whom the* inat *that made this book possible would have been sublimated into less wholesome pursuits.*

# Contents

| | | |
|---|---|---|
| Abbreviations | | viii |
| Introduction | | 1 |
| 1 | The Crucible of War | 21 |
| 2 | Projecting the Peasant on the World Stage | 41 |
| 3 | Reimagining the Nation | 61 |
| 4 | Between Aspirations and Limitations | 95 |
| 5 | Delegitimizing the Agrarian Alternative: The Diptych of Stamboliiski's Corruption and Radić's Treason | 121 |
| 6 | Drawing the Curtain | 153 |
| Notes | | 175 |
| Bibliography | | 201 |
| Index | | 226 |

# Abbreviations

| | |
|---|---|
| BANU | Bulgarian Agrarian National Union (*Bulgarski Zemedelski Naroden Suiuz* [BZNS]) |
| BAS | Bulgarian Academy of Sciences |
| BKP | Bulgarian Communist Party—Narrow Socialists (*Bulgarska komunisticheska partiia—tesni sotsialisti*) |
| CPPP | Croat People's Peasant Party (*Hrvatska Pučka Seljačka Stranka* [HPSS]) |
| CPY | Communist Party of Yugoslavia (*Komunistička Partija Jugoslavije*) |
| CRPP | Croat Republican Peasant Party (*Hrvatska Republikanska Seljačka Stranka* [HRSS]) |
| ECCI | Executive Committee of the Communist International |
| HSS | *Hrvatska seljačka stranka* (Croatian Peasant Party [CPP]) |
| IMRO | Internal Macedonian Revolutionary Organization (*Vutreshna Makedonska Revoliutsionna Organizatsiia*) |
| IPC | International Peasant Council, the executive organ of the Moscow-based Peasant International (*Krestianskii Internatsional: Krestintern*) |
| KPN | Confederation for an Independent Poland (*Konfederacja Polski Niepodległej*) |
| MAB | International Agrarian Bureau, the Green International (*Mezinarodni Agrarni Bureau*) |
| NRS | National Radical Party (*Narodna Radikalna Stranka*) |
| NSS | People's Peasant Party (*Narodna Seliačka Stranka*) |
| RPCC | Republican Party of the Czechoslovak Countryside (*Republikánská Strana Československého Venkova* [RSČV]) |
| UA | Union of Agriculturalists (*Savez Zemljoradnika* [SZ]) |

# Introduction

Lest there be any misunderstanding, the image on the cover of this book does not depict a snooty capitalist eyeing warily the bedraggled figure of a peasant that has so impudently interrupted his promenade. Aware of the possible misreading, the contributor of this image felt the need to replace the original title of the work with an awkward description that is supposed to fill in the knowledge gaps in the late twentieth- and early twenty-first-century Western audience. It reads, "The spectre of famine approaches the well-fed Aleksandar Stamboliyski the prime minister of Bulgaria in 1919 during the widespread European famine that followed in the wake of the First World War."[1] The creation date of January 1, 1900, that is listed in the details provided by Getty Images is obviously false, given that the image bears a date of 1920. Yet aside from the name of the contributor, Michael Nicholson, and the fact that the image entered the catalog of Getty Images via the Corbis collection, no other information regarding authorship or provenance is available.

Almost a century after this image was produced, the only data about this work that is unproblematic and indisputable is that it is available for licensing for a price ranging from 175 to 499 dollars, depending on the size of the image requested. In other words, the "value" that is attached to it and that has led to its archival preservation is just commercial. Historians are well aware of the contingency and selectivity in the archival record. Even within a properly catalogued and contextualized environment, documents do not "speak" by themselves, but require analysis and interpretation. The way a document was produced, used, or preserved—the positive knowledge about it—is vital to reconstructing and understanding its meaning. That said, it is always better to have something than to have nothing, especially since depictions of Stamboliiski are limited in number and in their ease of access, Getty Images having only three. However, the obverse is just as true: what has been stripped from a document and the reasons for doing it—the negative side—can be just as valuable in explaining its significance. So, let us attempt just such a reconstruction of the significance of this image, for in its relationship to the ideas of agrarianism and modernity and the golden age of the European peasantry, it can operate as a microcosm for what this book attempts to do.

My attempt to get more details about the illustration from Getty Images was unsuccessful. The representative who attempted to track down any additional information could only confirm that the contributor, Michael Nicholson, had passed away in 2016 and that the transfer of data at the acquisition of the Corbis collection could have been incomplete.² I was told that there was no information as to what the initials A. G. could refer to in the illustration. At this point, two avenues of inquiry presented themselves: whether to trace the image back to the moment of its submission as a digital copy within the image databases or to attempt to locate the original. Strikingly, these two avenues terminate in two diametrically opposed understandings of the same illustration. The contradiction and distance between them map out what can be gained by the reinsertion of a peasant political subjectivity, agrarianism, into the history of the interwar period.

The former option is quickly exhausted when one traces the origins of Corbis Images, via Continuum Productions, to Interactive Home Systems, a company founded by Bill Gates in 1989. The aim of that business was to digitize and license artwork for delivery into consumers' homes via vectors such as interactive television and digital frames.³ The business model changed over the years to the acquisition of image collections or news photography agencies for the purpose of expanding the licensing capability until Corbis was sold in 2016 to a photo and media agency, Visual China Group. In a division of terrain, Visual China Group is the exclusive distributor of the Getty Images collection in China, while it licenses the Corbis Images collection to Getty Images outside of China. From this it becomes clear that the logic governing the preservation of the Stamboliiski illustration is one that emphasizes its ornamental function and, even more troublingly, that there is an incentive to obscure the origin of the digital copy in order to be able to maintain its revenue-generating capacity. If one were to remain on this plane of analysis, the history that this image suggests cannot move beyond the sterile and straightforward representation suggested in the above-quoted caption: there was widespread famine after the First World War in Europe and its specter comes to trouble the well-fed Alexander Stamboliiski, as it would presumably have done to other prime ministers. Anything beyond is just speculation, but it would be plausible to infer either that famine served as a reality check to the ambitions of post–First World War leaders or that the destructive legacy of the Great War continued to rear its head even after the cessation of hostilities. However, all this distracts from the fact that this is a beautiful picture—masterfully drawn, employing a modernist composition and palette—and that it would serve as wonderful decoration. It projects a "Whig interpretation" of history in which even were the viewer aware that Stamboliiski was the head of a radical peasant political movement, he could still be appropriated, commodified, and just referenced in the inevitable path toward the triumph of progress and liberalism, alternatives and dead-ends be damned.

Despite the hurdles thrown in the path of the second, source-based approach, its pursuit is rewarded not only with the insights produced by the uncovering of an obscured historical perspective, but also by the replacement of the appreciation of art as decoration with the function of art as a perspectival jolt, a vista onto the past as a foreign country. The process of excavation begins with the insight that the initials A. G. are actually A. Б., and, coupled with some local knowledge, that they refer to Aleksandur Bozhinov, the father of Bulgarian political caricature. The establishment of authorship, however, is insufficient to fill in the gaps of attribution. The caricature on the cover does not appear in any of the published albums and collections of Bozhinov from 1907 to 2017.[4] Without a definitive title, medium, place of publication, or even the intended audience, the meaning of this work must be reconstructed creatively, and that begins with situating the author in his times. Aleksandur Bozhinov was born in 1878, the year the modern Bulgarian state was founded, and he observed, commented on, and participated in public life as a prominent figure until his death in 1968. His career spanned multiple administrations and systemic changes in the political structure of the Bulgarian state, yet, despite the critical nature of his art, he remained integrated and relevant. He was sufficiently valued by the political elite he would skewer to be generally celebrated, but, at the same time, he was disruptive enough to be repressed. He was expelled from the School of Art that would later become the National Art Academy for drawing caricatures of his professors, government ministers, and other important figures. He was imprisoned in 1907 for offending the honor of the ruling prince (kniaz), Ferdinand (tsar after 1908). Even though he was not affiliated with a political party, in the interwar period he was close to Ferdinand's son, Tsar Boris, and so, after the communist takeover in 1944, he was tried by the People's Court and again briefly incarcerated. He was a member of the Bulgarian Academy of Sciences (BAS) from 1929 and was the director of the Institute for Graphic Arts there from 1952 to 1953. While his popularity during the communist period waned in favor of more politically engaged contemporaries such as Ilia Beshkov (a critical sympathizer to the BANU [Bulgarian Agrarian National Union]), his stature was connected to his consistent ability to speak truth to power.

The thoughtful criticism that Bozhinov produced is a testament to his engagement with his subjects. For example, in an interview, Bozhinov recounts an anecdote that reveals his complex relationship with Stamboliiski and the agrarian regime. When some of his cartoons of Stamboliiski were stopped from being printed, Bozhinov authored an objection that argued that no one has the right to exercise censorship over the press, and the ban on publication was lifted. Bozhinov celebrated his victory by illustrating Stamboliiski as a subdued bull, thinking that would provoke another ban, but he heard later that Stamboliiski had laughed a lot and written on the illustration, "to be printed!" Bozhinov does not remember if he drew Stamboliiski afterward, but one day he was summoned to the prime minister's office and Stamboliiski offered to send him first as a trade representative to Berlin (Bozhinov had studied in Munich) and then as a press attaché, both offers that Bozhinov declined.[5] This context reveals a complex relationship with Stamboliiski that is predicated on a degree of mutual respect. It also begins to shed light on the figure of Stamboliiski in the illustration on the cover.

While the caricature advances a representation of Stamboliiski as arrogant and out of touch with bitter reality, it acknowledges that agrarianism personified in Stamboliiski is modern, present, and deserving of recognition. "Peasants" are in power and there is also the implication that they can be the subjects of a constructive critique. The assimilation of the agrarians as present and modern is visible in the choice of clothes for Stamboliiski, who did dress this way, as well as in the confidence of his bearing. That depiction is overdetermined since it matched the agrarians' own self-representation. By paying attention to this representational convention, we, as modern viewers, are reminded of the reasons why, given the semiotic conventions of the contemporary reader, we are so quick to assume a rich capitalist confronting inequality trope. We have regressed to the premodern conventions of representing the peasantry as abject and have excised the relatively short window of time in which there was no doubt in the mind of the contemporary reader that this was the peasant agrarian leader, Stamboliiski.

The connection to modernity is further reinforced by the staging on the yellow ground, which is an unmistakable reference to the yellow cobblestones with which the center of Sofia was paved between 1907 and 1908. Even to the present-day observer familiar with the architecture of Sofia, these cobblestones continue to be a symbol of the administrative and cultural center of Bulgaria, but it was even more amplified at the turn of the twentieth century. Almost 60,000 square meters were covered with the yellow cobblestones that were imported from Austria-Hungary in order to push back the mud in the city and rank it among the civilized capitals of Europe. The apocryphal story is that the cobblestones were a present from Franz Ferdinand for the marriage of Tsar Ferdinand, whereas their importation from Budapest was actually a modernizing municipal project initiated by the mayor of Sofia, Martin Todorov. To this day, when someone is described to be "from the yellow cobblestones" it means that that someone is connected to the center of power.[6]

It is at this point that the second interpretation of the caricature that I have been elaborating begins to acquire even more layers of meaning from the perspective of hindsight. Nineteen years before the production of the *Wizard of Oz*, the yellow brick road in the center of Sofia that is so implicated with the modernization of Bulgaria was the site of an uneasy and multivalent confrontation. The path to modernity had not brought the promised prosperity, let alone distributed the meager gains equitably. Instead close to a decade of war had reduced the populace to the beggarly status represented by the emaciated figure. The agrarians undertook an alternative modernity as a corrective, but a combination of hubris, structural limitations, and outright violence truncated this experiment. Given the brutal repression of the agrarians and Stamboliiski's murder in 1923, this image acquires another valence—the association with mortality. All too often in the interwar period, death was peering over the shoulders of agrarian activists, and political commitments were a matter of life and death. The story of this overdetermined failure and suppression just begins to be implied by the image, but it is also a story of forgotten audacity without which understandings and representations of the interwar period are incomplete. In turn, the reinsertion of the agrarian component or context fundamentally alters the meaning and significance of the image.

In the end, the two readings refer to the same image. Within a context characterized by imperfect knowledge, the second analysis relates to and, to some degree, relies on the distortions of the first. Yet still, the reading of this image in the second instance has restored something that was originally unavailable. This restoration and the integration to meanings relevant in the present, this interweaving, illustrates, in a dress rehearsal, the ambitions of this book. By making things visible, the deep analysis of the caricature, then, captures in a microcosm the aspirations and limitations of the golden age of the European peasantry, which are explored in a similar vein in this book.

One of the greatest casualties of modernity has undoubtedly been the peasantry. In Europe, it was squeezed, eroded, and relegated to the dustbin of history, even as its image as the repository of the nation was appropriated from the Left, Right, and Center. The renewed interest in the study of the peasantry in the 1970s—which saw the creation of *The Journal of Peasant Studies*, included the work of Teodor Shanin, Eric Wolf, and James C. Scott, and was marked by the "rediscovery" of Alexander Chayanov's *Theory of the Peasant Economy* in 1966—sought to return agency to the peasantry ironically at the moment when the European peasantry had finally disappeared.[7] The linkage of the peasant subject to resistance was thus explored in Southeast Asia or the Third World where it still had radical potential.

The relatively quick way in which the destruction of most of the European peasantry occurred in the twentieth century, however, has unfortunately strengthened the teleological orthodoxy that argues that the preservation of the peasantry was incompatible with the modern development of Europe. Given the often brutal means by which the countryside and agriculture were placed in the service of urbanization and industrialization, the arguments that slate the peasantry as doomed to extinction ring hollow. The circularity of these arguments often hinges on nothing more than the smugness of the victor over the inevitable loss of the vanquished. Yet, the enormous condescension of posterity is not easy to overturn, and not via the route of counterfactuals. Historical inquiry obsessed by questions of peasant viability or the exact causes of its failure to survive cannot hope to advance analysis in any meaningful way outside the necessarily limiting and well-trodden narratives of doom. Akin to debates such as those over the assignment of blame for the origins of the First World War, this type of inquiry produces a self-referential insularity that remains trapped within an analytical mode that cannot extricate itself from the boundaries of its postulates.

At the height of his power in 1923, the head of the BANU, Alexander Stamboliiski, summed up the significance of his politics for European history in the following way: "Today there are only two interesting social experiments: the experiment of Lenin and my own."[8] Taking the aspirations reflected in the quotation above seriously and rescuing the agrarian project from the enormous condescension of posterity is the foundation of this monograph. Briefly, it is about unpacking and restoring the significance of the golden age of the European peasantry between the two world wars by focusing on the paradigmatic cases of Bulgaria, Yugoslavia, and Czechoslovakia.

This monograph is, above all, an attempt to suggest an alternative approach to the unsatisfying neglect of the European peasant subject in the first half of the twentieth century. The provocative designation of the period between the First and Second World Wars as the golden age of the European peasantry is intended, first, to rupture the premise that significance and achievement must be necessarily tied to sociopolitical success. Second, the designation of a golden age is made necessary by the undisputable construction of a peasant subjectivity in that period that entered mass politics via peasant and agrarian parties, made claims on the nation as an articulate citizenry, and sought, in some cases, to elaborate and institutionalize a more ethical modernity, a third road between capitalism and communism.

In insisting that the interwar period cannot be properly understood historically without taking peasant politics seriously, this book consequently is also predicated on the claim that the peasantry's competitors in the political arena during the interwar period could not ignore it. Although the constitutive role of the peasant question in the policies and practices of bourgeois parties, royalists, the Right, or communists is not systematically explored here, it remains a foundation which runs through this work. The structure of this foundation is most visible in the choices not to examine agrarianism sui generis either in its own historical context or in relation to present-day historiography. Attention to the broader context serves not only to enrich the reading of agrarianism and correct its unwarranted marginalization. Analytical complementarity also demands the localization and examination of those sites where an archaeology of the past grounded in agrarianism can inform or recast the current historiography of the interwar period. Thus, the present analysis of the golden age of the European peasantry moves beyond history as just a corrective to omissions, to history conceived as a reexamination of the past that can contribute to and inform broader and current debates on topics such as nationalism, corruption, modernity, or the interrelationship of democracy and populism.

The implication of this twofold re-contextualization of peasant politics in the interwar period is a methodological solution to the problem of synthesizing agrarianism above the level of national politics. This book situates itself within a literature and historiography that is at times staggering in its depth and at others astounding in its paucity. At the national level, it has been impacted by the relative fortunes of the agrarian parties during the communist period, with the most developed being that of Bulgaria[9] and the least developed, that of Serbia.[10] In other cases, the national question has generated a body of literature on the HSS, Stjepan Radić, and Vladko Maček in Croatia[11] or on Milan Hodža in Slovakia[12] that supplements the work in the Czech Republic.[13] As the work of John Bell, Mark Biondich, and Daniel Miller illustrate, these types of studies are reflected in the English language.[14] What is notably missing, however, is a comparative monograph or one of synthetic scope. German literature has attempted to tackle this through collective projects that produced the volume edited by Heinz Gollwitzer[15] or the recently completed project on *Agrarismus in Ostmitteleuropa* at the Europa-Universität Viadrina.[16] Although these are significant contributions, I position myself against the tendency to do structural descriptions, like the "constellation" of agrarianism or the binary of *Bauern Agrarismus/Aristokratischer Agrarismus* (Peasant Agrarianism/Aristocratic Agrarianism), whose broad strokes obscure differences. In the stead of a unified movement to be described in its variations (a classificatory approach which also plagues treatments of agrarianism through populism),[17] I focus on the particular sociopolitical conjunctures that allowed the peasant parties to stake out a space for themselves in national politics first and on the international stage second.

In order to better clarify the position of this book within the field, a few words are necessary about its relation to prior scholarship and about the sources upon which it is based. The scholarship on the agrarian parties that was produced after the Second World War in Bulgaria, Yugoslavia, and Czechoslovakia was affected by the postwar political culture and the relationship of the communist parties to the agrarian ones. While this scholarship is, in general, solid and valuable, the skewing that affects it as a result of the conditions of its creation isolates it at the national level. For example, in Bulgaria, the coup against Stamboliiski in 1923 facilitated the construction of an instrumentalized historical narrative during the communist period that co-opted Stamboliiski's legacy as an imperfect precursor to the communist People's Republic after 1944. Thus, at its inception after the Second World War, historical inquiry into agrarian politics in Bulgaria prioritized the radical program of Stamboliiski, downplayed the competition with the Communist Party, and skirted, when not outright attacking, the politics of the BANU between 1923 and the start of the Second World War. The fact that the Left wing of the BANU was preserved as an allied political party throughout the communist period in Bulgaria illustrates both why scholarship on a "bourgeois" political formation could develop with minimal hindrance and why this scholarship was useful for, and thus supported by, the cadres of the communist-affiliated BANU.

This narrative, however, is based on a vision of Bulgarian exceptionalism, which discourages inquiry into the vibrant international exchange of ideas and cooperation between agrarians. Consequently, topics such as the Green International receive tangential mention that ends at the claim that the idea for the international originated with Stamboliiski. While the archive of the Bulgarian Agrarian Union's Representation Abroad is one of the few well-preserved collections that detail the international dimension of Bulgarian agrarian politics, until now it has virtually remained untouched. The individual particularities of each country covered in this work are different, but the politics behind the scholarship has kept the question of agrarianism fractured and provincialized. For example, Croatian scholarship is dominated by Radić and the question of

Croatian independence; similar is the interest in Hodža in newly independent Slovakia, while in Serbia the *Savez Zemljoradnika* is a historical footnote treated in only one monograph; the situation in the Czech Republic is marginally better.

It is my contention that the only way to restore the relevance and significance of the interwar agrarian experiment is to change the basis of inquiry and to mine the relevant source material in a new way. This book, therefore, is an experiment in how the national historiographies can be stitched together to provide a whole that is greater than the individual parts. One current in the book accomplishes this qualitatively different integration at the level of international interaction between the agrarian parties and in relation to their international context, while another either introduces new archival material or reexamines previously consulted sources with questions that reinsert the study of agrarianism into contemporary theoretical debates and developments in the historiography.

The archival research which provides the basis for this study is a departure from the insularity of the national approach. It was conducted in the archives of four Eastern European countries and contains material from the Russian State Archives. Besides the challenge of gathering similar material in these institutions that could provide a basis for comparison, the hunt for the archive of the Green International, which took me through the territories of Bulgaria, the former Yugoslavia, and the former Czechoslovakia, and which could not be found and was most likely destroyed, meant that I tried to reconstitute it from the personal archives of various agrarian figures from each of these countries as well as from interior ministry records concerned with the monitoring and suppression of agrarian activity. Research of such magnitude and duration inevitably transformed the nature of my project as the discoveries and gaps in one country changed the agenda and the questions asked in the next. Thus, the nuanced and limited portrayal of agrarianism that this book proposes is a direct product of the necessity to coordinate and reconcile the specificities of the national developments with the particularities of the international stage.

Since the reconceptualization of agrarianism that underlies this work involves historiographical revisionism, its site is in the individual chapters where each instance of revisionism drives that chapter's argumentation and structure. In some instances, it calls to task certain conceptions and evaluations that are present in Eastern European historiography. Just as often, however, this revisionism operates through the introduction of theoretical conceptions present in historiographies and literatures that have not yet been incorporated in Eastern European studies. The challenges and nuancing I propose are always related back to a suitable and representative example of agrarian praxis. It is my hope and intention that this methodology of accomplishing a critique by means of the re-inclusion of a heretofore neglected and marginalized historical experience opens new directions of scholarship that can reinvigorate the field. It is a further conceit of mine that this approach has historiographical bearing on European history, in which Eastern European studies are relegated to the periphery. In this way I seek to do justice to the self-understanding of my agrarian subjects, who not only considered themselves European, but thought of their ideas and contributions not as an irrelevant footnote, but as a way to make Europe and the world a more equitable place.

This book defines agrarianism as the imagination and articulation of a peasant political subject. The struggle to give agency to the peasant as an articulate subject in the encounter with modernity is what sets apart the interwar period from the times in which the peasant has been conceived as a mute object to the forces and exigencies of modernity, inhabiting a parallel moral and economic universe. Furthermore, this definition captures the intentionality of a sociopolitical project at the moment of its inception and emphasizes the relational aspects of identity. Althusser's concept of "interpellation" is useful in grasping the thrust of this definition, namely that ideology is constitutive of subjectivity and that it is a representation of the imaginary relationship of individuals to their real conditions of existence.[18]

\* \* \*

This work is a complex structure that advances and illustrates its main arguments through a fragmentary process of accretion. Its chapters are facets of the general problem of agrarianism and each chapter is a different lens into that problematic. As experiments in rethinking, recasting, or recontextualizing, the chapters' structure is designed to prioritize the development of an argument over a strict adherence to a chronological exposition. The incorporation and close reading of key texts and polemical engagements also contribute to the difficulty of this work. It is imperative, therefore, to present the main argumentative structure of the book as it is developed through the chapters and within each individual one from the outset so that the detail within the chapters can be better appreciated.

The thematic chapters of this book alternate between being strategic probes of evocative micro histories and being broader theoretically informed overviews in order to illustrate and clarify the analytical frame. The most radical expression of interwar agrarianism, that of Bulgaria, and the man responsible for it, Alexander Stamboliiski, serve as the center of this book.

The juxtaposition of this center to the development of agrarianism in Czechoslovakia and Yugoslavia, as well as to various oppositional formations such as the Peasant International in Moscow allows this research to overcome the national parochialism that has contributed to the sidelining of the study of agrarianism. The innovative structure of the work is above all a demonstration of the rich possibilities still open to researchers in this field to reinsert the study of agrarianism into contemporary theoretical debates and developments in the historiography. The book explicitly engages agrarianism with the theoretical literature on nationalism, corruption, the subaltern, as well as making possible the connection to the problematics of modernity, politics as systemic change, and transnational and global history.

The function of the foregrounding Chapter 1, "The Crucible of War," is to advance the argument that the agrarian movements of Bulgaria, Yugoslavia, and Czechoslovakia were forged in the revolutionary formation and transformation of these countries during the First World War. The socioeconomic and political impact of the war was a unifying experience that meant upheavals across the board and novel national and institutional construction in its aftermath. In all cases, the agrarian parties of these countries were propelled to the forefront of national politics. In comparison to their marginal positions prior to the war, apart from the *Savez Zemljoradnika* in Serbia, which was formed immediately after the war, these parties were confronted with a radically transformed sociopolitical landscape in which prewar forces and institutions were discredited, severely weakened, or even expelled.

The war was not just a transformative context that left these parties untouched. On a second plane of commonality, the experience of war radically transformed the parties themselves. The articulation of a vision for the postwar order coupled with an enormous expansion of the parties' base meant organizational and programmatic change so that these parties could adapt to their role as a major actor, if not the principal one, on the national stage. Further, the consequence of mass politics in the demographic context of half to three-quarters of the population being engaged in agriculture produced a legitimacy and urgency in the parties based on the recognition that the time of the peasant as a subject had arrived and could not be squandered.

While the war defined the central position of the agrarian parties in the postwar period, the individual positions that they adopted and the trajectories they took produced a rich variation. The first chapter thus introduces the proposition that the four autonomous peasant movements that emerged out of the ashes of the First World War represent three faces of the alternative modernity that was articulated in the golden age of the European peasantry: agrarian radicalism in Bulgaria, the peasant nation in Yugoslavia, and centrist agrarianism as the guarantor of parliamentary stability in Czechoslovakia.

The momentous transformations in the crucible of war in the Balkans spanned nearly a decade given the foreground of the two Balkan Wars. The central component of this chapter, which details the anti-war position adopted by the BANU and Alexander Stamboliiski, is the treatment of the transformation of that position from a liability to an asset given Bulgaria's series of "national catastrophes." Opposition by the BANU in the war years and repression against it produced ideological clarification, republican radicalization, and provided it, alongside the Communist Party, with a monopoly on legitimacy when the traditional political forces were discredited in the defeat. The enormous numerical growth of the party in 1919 could permit its treatment as a novel political formation, although that had to be tempered by the compelling case in favor of continuity given by its consistent praxis in the second decade of the twentieth century.

The hegemonic position of the BANU at the start of the interwar period was a direct consequence of defeat and the imperative for rupture with the past and a delineation of a new course; its radical attempt to reorder the Bulgarian polity was framed by this condition and was both enabled and limited by the status of Bulgaria as a defeated country. The ruptures engendered by the war, which were the universal sine qua non for the creation of space for peasant politics, encompassed a variability that was expressed differently in the case of "victor" nations. In Yugoslavia, the newly formed *Savez Zemljoradnika* that had its base in Serbia could capitalize on the desire for land reform and a critique of the NRS (*Narodna Radikalna Stranka*), which had departed from its socialist and peasantist roots in the nineteenth century. Through the incorporation of other recently formed regional peasant parties such as the *Savez Težaka u Bosni i Hercegovini* (Bosnia and Herzegovina), *Težački Savez* (Dalmatia), and the *Seljački Savez* (Croatia), the *Savez Zemljoradnika* strove to form a Yugoslav peasant platform. In the first years of potentiality after the war, the *Savez Zemljoradnika* had its largest electoral successes. In the long run, it could neither supplant the dominance of Nicola Pašić's NRS in Serbia nor could it challenge the democratic peasant nationalism of Stjepan Radić's CPPP (*Hrvatska Pučka Seljačka Stranka*) in Croatia. Nonetheless, it succeeded in articulating and occupying an alternative political space to the nationalist politics in interwar Yugoslavia.

The war also transformed the CPPP that had around 15,000 members in 1914 into the preeminent party of Croatian nationalism, the *Hrvatska Republikanska*

*Seljačka Stranka* (CRPP), with a membership exceeding 1 million in 1921. This transformation, the insistence on Croatian rights, and the opposition to Belgrade set the foundation for Radić's party to become the dominant political force in Croatia in the interwar period. For both the *Savez Zemljoradnika* and the CRPP, the war was not only an opportunity that projected them on the national stage but also a limitation, given Serbia's victory. The initial social radicalism had to give way in the face of the resilience and dominance of Belgrade-centered nation building.

In the newly formed Czechoslovakia after the First World War, the RSČV had gained greatly in strength from its less than 100,000 members in 1914. Even after its merger with the Slovak *Slovenská národná a rolnícka strana* (Slovak National and Peasant Party), its electoral strength was initially limited. Nonetheless, the centrist positions it adopted secured it a key role in the moderate parliamentary coalitions during the postwar revolutionary moment. This role was instrumental to the institutionalization of the "politics of compromise" in the First Republic and secured the position of the party as the guarantor of that order. In order to illustrate the transformed position of the agrarian parties after the First World War, Chapter 1 concludes with a brief survey of the prewar origins of these agrarian organizations.

Chapter 2, "Projecting the Peasant on the World Stage," contextualizes the projection of agrarianism on the world stage through its organization, the MAB (International Agrarian Bureau). This chapter is not an institutional history of the organization. Although this book project began as a hunt for the archive of that organization and, subsequently, when it could not be found, gathered as many extant traces of its correspondence in the organizational fonds of the agrarian parties and the personal fonds of its leadership in Bulgaria, Serbia, Croatia, and the Czech Republic, this material is insufficient to reconstruct the inner workings of the organization. At its best, it can periodically annotate and correct the self-representation of that organization through its official periodical, the *Bulletin Mezinárodního Agrárního Bureau*. For reasons related not only to the dearth of material, to its negative evaluation by its communist-affiliated competitor, the Moscow-based IPC (International Peasant Council), better known as the Red Peasant International (*Krestintern*), to its designation as a reactionary bourgeois institution during the communist period after the Second World War that even generated show trials against its "agents," but also to the scholarly neglect that has plagued the study of European agrarianism outside certain well-trodden aspects of national history, the historiography that touches upon this organization is plagued by mystification.

Two monographs characterize the range of approaches and, incidentally, they are the only works of some length on the subject, although the second one devotes only half a chapter on the subject and a bit more in passim. The 1967 monograph of Maksim Goranovich, whose title translates to *The Collapse of the Green International*, is an ideological polemic against the organization that he portrays as a reactionary formation that cleared the way for fascism.[19] The antecedents to this line of argumentation lie in the polemical battles between the *Krestintern* and the MAB that intensified in the mid-1920s. The dismissal of the Green International, as the bureau was more commonly known, was motivated by the perceived hindrance it could exert on the Communist International's (*Comintern*) strategy to penetrate the village and organize the peasantry in alliance with the workers toward opening a new front in the world revolution after the post–First World War revolutionary moment failed to produce the expected soviet republics. The focus on top-down ideological polemic causes Goranovich to overemphasize the executive power of the organization in its representation as an instrument of class enmity.

While the Soviet scholarly orthodoxy that dismissed the interwar agrarian rivals depends on the question of how a reactionary organization could be robust enough to hinder the logical course of historical materialism, scholarship from across the Iron Curtain inverted the question to focus on the weakness and failure of the organization and agrarianism in general. George Jackson's nearly contemporaneous monograph to Goranovich's from 1966, *Comintern and Peasant in East Europe 1919–1930*, offers an opposite distortion.[20] While the book's chief focus is on the *Krestintern*, its cursory treatment of the MAB judges both organizations by the same yardstick. The Green International, according to Jackson, neither had the resources of a revolutionary state apparatus behind it nor the organizational discipline of the communist parties nationally and internationally in the way the *Krestintern* did. Its aim was not to coordinate revolution but, through example and information exchange, to empower the national agrarian movements that composed its membership. Yet in the zero-sum logic of winners and losers, the question of failure is overemphasized. While I believe much of this is related to a Cold War mentality that projects the onus of failure on agrarianism for its inability to prevent the communization of Eastern Europe after the Second World War, fixating on agrarianism's "failure" to impose its maximalist program during the interwar period due to internal ideological and organizational factors produces a reductive de-emphasis of the contextual difficulties that brought even liberal democracies to their knees at the time. Incidentally, the *Krestintern*, with its logistical and organizational superiority, dissolved itself in 1931, while the disbanding of the MAB followed the Anschluss of Austria in 1938 and the demise of Czechoslovakia in 1939.

As a counterpoint to this scholarship, my intention in Chapter 2 is to foreground the potentialities and the alternative developmental streams that contextualize the founding of the MAB. A more detailed treatment of the activities of the organization follows in Chapter 4, and it reappears again in Chapter 6. The second chapter begins with an appropriation of the idea of the Green International in the pages of the *New York Times* from 1921 that is indicative of an effort to tame and channel the agrarian challenge. This section begins a layering of three imaginaries that frame the function of the MAB, the second being the 1927 internal Czechoslovak rewriting of the history of the formative years of the organization, and the third, the expectations toward that organization by one of its founders, Alexander Stamboliiski.

The chapter then argues that the history of the crystallization of the MAB in 1923 is explained by multiple currents in the development and implementation of agrarianism and that two frames are useful for contextualizing that history—namely, the federalist idea itself and the competition with the *Krestintern*. Further, the international stage, despite the hopes and efforts of the agrarian parties to recast it, was severely constrained by the Versailles system. The chapter examines Stamboliiski's efforts to soften Bulgaria's isolation after the Paris Peace Treaties. A survey of the federative idea in Southeast Europe frames Stamboliiski's proposal for a South Slav federation. In turn, the rebuffing of Stamboliiski's diplomatic efforts contextualizes the birth of the MAB as the pursuit of that idea by other means. While the coup against the BANU and the death of Stamboliiski in 1923 submerged this course in the interwar period, this section completes the circle by discussing the 1942 revival of the federative idea by Milan Hodža, one of the Czechoslovak agrarian founders of the Green International and prime minister of the country from 1935 to 1938.

The chapter then focuses on the other frame that contextualized the formation and activities of the Green International—namely, the *Krestintern*—and provides a survey of that institution. The chapter concludes with the insistence that once the divergent trajectories represented by Czechoslovakia and Bulgaria resolved into the truncation of the decisive Bulgarian input, the MAB developed according to the directives and interests of the Czechoslovak Republican Party. The ensuing recasting of the institution in the 1920s and 1930s can be better explained by focusing on the configurations of the international situation, as well as on the competition with the *Krestintern*, rather than on the questions of *whether* or *why* the institution was a "failure."

Chapter 3, "Reimagining the Nation," introduces the practice of alternating chapters between case study analysis and broader synthetic work. The agrarian movements of Bulgaria and Croatia in the interwar period receive a very different treatment in the literature. Whereas the Bulgarian case has been described as anti-national(ist), the Croatian is represented as the embodiment of a national movement. This polarity, however, ossifies and essentializes the complex relationship of the agrarian movements with nationalism and the imagined community at a moment when the character of that community was being transformed through the introduction of the peasant subject. The first part of this chapter illustrates how the different outward expressions of nationalism in the Bulgarian and Croatian agrarian movements are in fact based on very similar conceptions of the peasant rights and the peasant state.

At the heart of this chapter is a case study of the conflict over the implementation of the BANU's orthographic reform in 1921. This microhistory is wrapped in several layers of analysis that contextualize this episode and permit its relation to several currents in nationalism studies. The chapter traces the correspondence between the conjunctural chronology of changes in Bulgarian orthography and the mutable nature of Bulgarian nationalism by opposing this inquiry to the continuing influence of the civic/ethnic antinomy in the scholarly literature of Eastern Europe. Showing that orthographic reform was an important element of the discourse of ethnic as well as civic nationalism not only fractures notions of structural continuity predetermined by the path to a nation-state but is exemplary of the fluidity of national identity as it interacts with parallel discourses of modernity. The orthographic reform reimagined the national community and in a significant, structural way paralleled a shift in the organization of the nation. It was intended to and eventually succeeded in recasting the role of the Bulgarian citizen. The detailed micro study of the language question, culminating in the orthography debate of the 1920s, proposes that the spelling reform was much more than a matter of sociolinguistics. It marked a fundamental change in the Bulgarian polity and was associated with a moment of restructuring, democratization, and reshuffling of elites, the rearrangement of norms that favored different social and cultural constellations. In other words, the picture of nationalism that it elicits is one that intimately grapples with one of the most central political processes of the modern Bulgarian state—the question of democracy and popular sovereignty. Theoretically, this second part elaborates an opening in the treatment of Eastern European nationalism that draws inspiration from the work of the Subaltern Studies Group. While a strict transposition of concepts is counterproductive given the contextual differences, the critical juxtaposition of Eastern European agrarian nationalism to the emancipatory project of subaltern studies facilitates the extraction of the former from the stale narratives that obscure and suppress the radical agrarian merger of the nation, the peasant state, and the peasant subject.

Chapter 4, "Between Aspirations and Limitations," pulls the lens back out again. This chapter begins with a focus on the BANU's Representation Abroad in Czechoslovakia after the coup d'état against the Stamboliiski regime in 1923. It relates its activities to the politics of the Republican Party that was hosting it in Czechoslovakia. In exile, the Representation Abroad represents one of the lowest points in the political fortunes of agrarianism in the interwar period. Yet my analysis aims to show the significance and potentialities even there. The Representation Abroad's biggest achievement was that at a time of seeming hopelessness that anything could be done to restore political normalcy and a semblance of democracy in Bulgaria, its members tenaciously kept the spark alive for themselves, the beleaguered BANU, their agrarian partners in the MAB, and in front of European public opinion.

I illustrate this inspiration by examining an instance of history from below which involved the dispersed Bulgarian agrarian émigrés in Europe creating the Union of the Bulgarian United (*zdruzheni*) Agrarians Abroad and generating a critical resolution against the internecine conflicts in the BANU. This episode illustrates my definition of agrarianism as not only the expression but also the establishment of a moral economy in the peasant that altered his relation to modernity from a transitory object into an agent with a stake in its constitution. I counterpose this approach to agrarianism to what I see as an excessive focus on ideology in the study of agrarianism.

As a result, my critique is differentiated from the classificatory approach to agrarianism in classical studies of populism as well as the recent effort to synthesize agrarianism in East-Central Europe that is hindered by a structuralist frame. While I argue that the fate of agrarianism was context specific and that it produced the three faces of agrarianism in the national contexts that are the focus of this monograph, I still propose three universal initiatives that underwrite the agrarian project: parliamentarism, land reform, and the cooperative movement. As an illustration, this chapter then illuminates these initiatives in the radical agrarian phase of Bulgarian history. The chapter concludes with a summary of the activities of the Green International in the first three years of its existence. The merit of this exercise in presenting the work in the *Bulletin of the MAB* lies not only in the fact that it has never been done before. Presenting its work in this way shows the varied initiatives and coverage it engaged in and in this way corrects the various holes, misrepresentations, and errors in the scant literature on the subject.

Chapter 5, "Delegitimizing the Agrarian Alternative: The Diptych of Stamboliiski's Corruption and Radić's Treason," discusses the delegitimation campaigns against the agrarian movements in Bulgaria and Croatia. This chapter returns to the method of case study investigation that is then layered in theoretical analysis. As in Chapter 3, the resonance of neglected moments of agrarian history in contemporary scholarship justifies the importance of this history.

By manufacturing a posthumous charge of venal corruption against Stamboliiski, Alexander Tsankov's regime attempted to fundamentally discredit and erase the systemic alternative offered by the reforms of the BANU up to 1923. I reconstruct a detailed micro-history of the corruption trial against Stamboliiski based on heretofore unutilized archival sources, not simply in order to correct a historical misrepresentation. Rather, I am interested in corruption as drama: the way charges are articulated, how they are directed, and the way they ultimately operate as a legitimizing tool in the context of systemic transition. The instrumentalization of corruption for political ends lends itself particularly well to the extension of vices from the personal to the systemic level. Analyzing the growing theoretical literature on corruption, this chapter reaches the conclusion that the obsession of the capitalist system with the particular type of corruption that involves personal pecuniary enrichment is far from coincidental and is informed by the logic of capital accumulation within a free market economy. According to this logic, the private drive to amass is best protected from degenerating into corruption within the liberal market democracy. Therefore, it is imperative to delegitimize any alternative systems whose aim, at the very least, ideologically is to excise that type of self-interest in the first place.

The second case study looks at the charges of treason brought about by Radić's trip to Moscow and the entry of the CRPP into the *Krestintern*. The delegitimation campaign against Radić was designed to weaken the organizational strength and political prestige of the CRPP through the misrepresented association with the communist menace. It was essentially a blackmailing operation, a maneuver to force a differentiation between communism and agrarianism that, at its extreme, might cause the abandonment of the republican idea and the recognition of monarchy. Terror and revolution were the monsters against which the interwar capitalist order was trying to protect itself. It succeeded in outlawing the weaker communist parties in Bulgaria and Yugoslavia but was forced to accommodate the agrarian parties, albeit under various degrees of repression.

The concluding chapter, "Drawing the Curtain," tackles the problem of the denouement of the agrarian moment during the Second World War and the years immediately following its conclusion. It begins in a historiographic mode that challenges the simplified model of the elimination of the peasantry as a political subject in the West, through the irresistibility of capitalist logic, and in the East, through Stalinist collectivization and terror, and calls to task the exclusion of agrarianism at the level of the survey literature in the Eastern European field. In marking the end of the agrarian alternative, this chapter refuses to locate it in a "tragedy" of communization. The real tragedy, it argues, began with the separation of the agrarian political elite from its base during and after the Second World War and is complete now, in the post-socialist period, in the grotesque revival of agrarian parties that struggle to speak for and exploit the memory of a peasantry that no longer exists. In marking the Second World War as the revolutionary and transformative moment that took away the relevance of agrarian politics, this chapter complements the treatment of the First World War as having created the conditions for the initial elaboration of agrarian politics. Touching on recent work on collectivization, this chapter normalizes the socialist experience and proposes that the peasantry, having already constituted itself as a political subject in the interwar period, was transformed as much as it transformed itself in the encounter with state socialism.

To mark the erosion of agrarian relevance, this chapter contrasts the program of the BANU from 1919 to that of BANU "Aleksandur Stamboliiski" from 2008. In a similar mode, it juxtaposes the pathos of Vladko Maček in 1949, when he provided a characterization of the International Peasant Union while acting as its vice president, to the even more dismal politics of the *Hrvatska seljačka stranka* that was "reconstituted" in 1990. A look at the post-socialist developments in Serbia, the Czech Republic, and Slovakia completes the picture.

Finally, this chapter examines the experience of Czechoslovakia in the years surrounding the Second World War in order to explain the fate of the Republican Party. It is a paradigmatic case of how the Second World War changed the terrain to the exclusion of agrarianism years before the communist takeover in 1948. Thus, when this chapter concludes with a description of the show trials against the agents of the Green International in 1952, it underscores the absurd element of the mobilization of an offensive against an already broken "foe."

# 1

# The Crucible of War

At the height of its power, yet barely two-and-a-half months before the coup d'état which violently brought it down in the summer of 1923, the majority government of the BANU was still referring to the legacies of concussion and dislocation brought about by the First World War in order to frame policy. The official organ of the BANU proclaimed:

> The world war caused an economic crisis in the life of countries, worsened international ties, tripped up the cultural progress of humankind, results equally bitter to victors and [the] vanquished. . . . The situation is tense for all, be they Germans, French, Turks, Serbs, or Bulgarians, and united they cry out, under the pressure of social needs, for an end to murderous wars and economic slavery. The specter of a war, which will complete the destruction of the world, scares equally all nations; that is why lately there is talk of concessions and understanding in order in this manner to avoid new social tremors in international relations and to guarantee and strengthen the existing economic and agricultural life of individual nations. This is especially so for Bulgaria.[1]

The rhetorical recourse to the trauma of the war should not be seen only as a means to an end in the political vocabulary of the agrarian movement; it articulated the resultant processes that propelled the agrarian movements of Eastern Europe to the center of the political stage in the postwar settlement. The First World War radically transformed the continent, and, as such, the experience of Bulgaria, the Triunine Kingdom, and Czechoslovakia is firmly imbedded within the larger European experience. Yet at the same time, the unique emergence of a powerful, Center-Left agrarian moment requires a situating within the general European context that illuminates not only the successes but also the limitations of the agrarian movements to harness the legacy of the First World War to radical projects of social transformation and nation building. Only through the lens of the First World War and the immediate postwar years do the momentous transformations in these three countries become intelligible. They include the transformation of the BANU into a pluralistic force for social revolution, the conversion of the *Hrvatska Republikanska Seljačka Stranka*, HRSS (Croat Republican Peasant Party, CRPP) beyond a party into the movement of the entire Croat people,[2] the elevation of the *Republikánská Strana Československého Venkova* (Republican Party of the Czechoslovak Countryside, RSČV) into a keystone of Czechoslovak parliamentary order, and the emergence of the *Savez Zemljoradnika*, SZ (Union of Agriculturalists) as a challenger to the Radical Party's hegemony in the Serbian village.

If European enthusiasm at the start of the First World War was mobilized toward national unity and the elimination of social tensions, this had evaporated by the war's end and set the stage for social and national revolution. The contributing factors were not restricted only to the enormous human and economic losses caused by the waging of total war. Decisive as well was labor discontent, for the mobilization of the workforce into the war machine resulted in not only the increased bargaining power of labor unions and the transformation of labor markets and social institutions that guaranteed rights to strike, the eight-hour workday, and eventually significant expansions of welfare policy, but also labor militancy in the last years of the war that flowed into the postwar period. Demobilization strengthened the calls of returning soldiers for a voice in the policies of the state, and this put enormous pressure for land redistribution when combined with the demands from the agricultural sector and the village. At the level of state and nation building, the redrawing of the map of Europe, particularly in the lands formerly held by the Habsburg, Romanov, and Ottoman Empires, bequeathed upon the interwar period a difficult calculus pitting national self-determination against territorial integrity.

The vast literature on the cultural memory of the First World War underscores the disillusionment and cultural rejection that set a generation apart. But, more often than not, in the short term, this meant either a discrediting or marginalization of the traditional political forces that had been implicated in national catastrophes for the defeated states or else the senseless slaughter that engulfed the victors as well. One could then argue that the experience in the First World War, which had brought down the very system that it had sought to maintain and strengthen, was transformed into the cautionary tale that underwrote the efforts to ensure stability in the early 1920s.[3]

In trying to overcome the shortcomings of scholarship encumbered by "the habit of thinking in terms of national quarrels," Fritz Fellner has sought a corrective through emphasis on continuities.[4]

> Perhaps the triumph of the principle of continuity in social and financial matters was largely due to the fact that the overthrow of the monarchy in the autumn of 1918 was not a revolution but only a national and constitutional upheaval. Apart from the brief Communist rule in Hungary, the existing social and economic order was preserved in the former provinces of the Habsburg monarchy, and, what was still more important, the bureaucracy and the administrative apparatus remained intact. The members of the aristocracy who had occupied the key positions in government until the final years of the war were replaced by the leaders of national mass parties, but under the supervision of these new ministers the old administrative apparatus continued to operate in all the Successor States. The revolution that aroused so much fear, the specter of which appeared so threatening to the statesmen assembled in Paris, never occurred in Central Europe.[5]

In Fellner's argument, one can see an antecedent to the crucial scholarship of Charles Maier in *Recasting Bourgeois Europe*, yet, it is precisely this work that complicates facile distinctions between rupture and continuity.[6] The political space opened up by the First World War introduced for the first time agrarian political actors on the national stage that had either direct or potential access to the transformation of their societies. Within the limits imposed by the particularities of each country, these agrarian formations provided a buffer against the expansion of social revolution along the lines of the October Revolution. In this sense, they served as a buffer to socialist or communist expansion. But this holds true only in conjunction with the statement that their role as a buffer went hand in hand with their own programs for fundamental transformation. To put this differently, the claims advanced by the agrarian parties at the very least articulated a political agenda that had to be taken seriously and either incorporated or confronted by any rival political entity in the interwar period. This would describe the position of the *Savez Zemljoradnika* or the RPCC, even if one were to retrench around the myopic refusal to comprehend the unique parliamentary stability of Czechoslovakia throughout the interwar period, of which the latter party was a guarantor, as a revolutionary achievement. As far as the *Naroden Suiuz* or the *Seljačka Stranka* are concerned, however, there is no other way to describe their activity but as politics in a new key.

While the argument so far has emphasized the commonalities in the transformative experience and legacy of the First World War, attention to the different ways in which political space was occupied by the agrarian parties in no way undercuts the broader framework. In what follows, a history is presented of the trajectory to prominence that each agrarian party took during and immediately after the war. To temper the teleology inherent in such a reading, special attention will be paid to the way the war and its aftermath effectuated transformations in these agrarian political parties themselves.

## The war decade and the BANU

In the case of Bulgaria, the experience of the First World War is intimately tied to the Balkan Wars, and not only insofar as it offered a possibility of reversing the territorial "losses" incurred in the Second Balkan War. The casualties in the First World War, which amount to 100,000 killed and 200,000 wounded came on top of the 58,000 killed and 105,000 wounded in the Balkan Wars.[7] The territorial gains made in the First Balkan War were lost in the second, including Southern Dobrudzha, Eastern Thrace, and most of Macedonia. Western Thrace, which had been retained with much difficulty at the Treaty of Bucharest was lost following the First World War at the Treaty of Neuilly-sur-Seine. Between 1915 and 1919, the grain harvest fell by almost half, while inflation between 1900 and the beginning of 1919 stood at 1,100 percent.[8] The national catastrophe discredited both the bourgeois parties and the throne, Tsar Ferdinand being forced to abdicate in favor of his son Boris in October 1918, but the rise of the Agrarian Union was not due to simply occupying a power vacuum.

Already from the run up to the First Balkan Wars, the Agrarian Union had staked an anti-war stance. Alexander Stamboliiski and the BANU MP Stoian Omarchevski, who would later become Stamboliiski's Minister of Education, made the case in the party organ, *Zemledelsko Zname*, that the question of Macedonia was used to distract from domestic social reforms and that "the real enemies of the Bulgarian people . . . were absolutism, reactionary social policies, and blind nationalism, against which all peoples, Bulgarians, Turks, Serbs, and Greeks, should unite."[9] Military censorship, which came into effect with mobilization, deprived the agrarians of their tribune as *Zemledelsko Zname* was forced to shut down.

The perorational character of Stamboliiski is visible in two confrontational "meetings" that he had with Tsar Ferdinand at the immediate start and end of the First World War. These conversations are worth examining in greater detail as they not only bookend the war and reveal Stamboliiski's position vis-à-vis the conflict, but also reflect the changed position and authority of the BANU by the end of the war. The meetings were first published and presented with a commentary by Nikola Petkov, the former editor of *Zemledelsko Zname*, who would take over the leadership of BANU "Alexander Stamboliiski" during the Second World War and would later mount the most serious resistance to the communization of Bulgaria after 1944.[10]

On August 28, 1915, driven by the rumors that Bulgaria's entry in the war on the side of the Central powers was imminent, a delegation of the agrarian parliamentary group, composed of Alexander Stamboliiski, Dimitur Dragiev, and Aleksandur Dimitrov, sought to dissuade the prime minister, Vasil Radoslavov, from this course.[11] Despite the arguments, the delegation was sent away after being told that everything already has been decided and that there would be no turning back. Suspecting that this decision had been made at the court, a meeting was called of all the parliamentary oppositional groups (Broad Socialist, Narrow Socialist, Democratic, Radical Democratic, Progressive Liberal—Danevist, National—Geshovist, and the BANU) to determine a course of action. It should be remembered that the BANU was the second largest party in parliament after the November 24, 1913, elections, which had given it forty-eight deputies and 20.9 percent of the vote to the ninety-five deputies and 38.2 percent of the vote for the Liberal Coalition. As I have already argued, the crisis of the Balkan Wars links to that of the First World War in the case of Bulgaria, and, as John Bell has written in discussing the 1913 electoral results, "this swing of the Bulgarian electorate away from the established parties [is] a movement that Bulgaria's defeat in the First World War was to complete."[12] Based on the BANU's oppositional strength and its consistent condemnation of the First World War already from 1914, the meeting of the oppositional groups tasked Stamboliiski to lead representatives of most of the groups to an audience with the king on September 4, 1915.

Stamboliiski's speech to the king was at first a very daring exposition of the arguments against entry into the First World War. Given the specific Bulgarian context, Stamboliiski summed up the planned entry into the war as baseless adventurism with no hope of success. He distinguished the Tsar's adventurism from popular national action to stress its complete unrealizability: "But every adventurism, when it has the essential element for its broadest realization, i.e. when it is a popular and attractive action to the national masses, it achieves the unassailable epithet 'Народно дело' [patriotic act] and can bravely present itself at the international market and can bravely defend its correct or incorrect cause."[13] Assuming the mantle of a tribune for the Bulgarian people, Stamboliiski enumerated five characteristics of the Bulgarian people:

1. Its feelings toward Russia have not disappeared. That is sad, but a fact. It is sad because it limits the free rein of the Bulgarian statesman.
2. Its horrible impression of the destruction it survived [the Second Balkan War] also has not disappeared.
3. Its faith in the rulers is dead.
4. The fear of war, and especially an unpopular war, and at that a war on several fronts, is a terrible nightmare on its consciousness and constantly stifles it.
5. And most importantly, its faith in You, Your Highness, is completely shaken and killed. In its eyes, in the eyes of the nation, after the catastrophe of 16 June 1913, you have been stripped of the renown of being a fine diplomat.[14]

If the king were to persist with his course, Stamboliiski warned him of the consequences:

> In 1913, immediately after the demobilization of the army, we began to receive letters, delegations from all corners of Bulgaria began to appear before us, through which the bitter and dishonored Bulgarian people implored us to lead the struggle *for the quick and decisive finding and punishing of* **those culpable for the catastrophe, of which, first and foremost, are You.** Indeed, those gentlemen around you wholly attempt to elude responsibility, but we have never separated them from you. And well, since we believed then that *the catastrophe was not a malicious action by you and your* ministers, and since we were afraid about the fate of Bulgaria from the Romanian invasion, we decisively positioned ourselves against that brutal popular feeling and we succeeded in diverting it. *Remember though, that if tomorrow you perpetrate* **the same criminal act,** we, the people of the Agrarian Union, not only will not halt the national resentment against You, **but we, ourselves, will become its expression, when we serve You its harsh but just sentence.**[15]

The audience did not end well in two respects. First, when Stamboliiski reiterated that the king's course would lead to the loss of his head, Ferdinand replied, "Don't worry about my head, I am old, think about yours, as you are still young."[16] Indeed, on September 13, 1915, a warrant for Stamboliiski's arrest was issued and he was incarcerated on September 15, 1915.[17] Second, Stamboliiski's intervention in no way halted Bulgaria's entry in the war. The day after the meeting, Radoslavov signed the Treaty of Friendship and Alliance between Bulgaria and Germany, thus placing the country on the side of the Central powers. Mobilization followed on September 21, and Bulgarian troops moved into Serbia on October 11, 1915.

A military court initially sentenced Stamboliiski to death for treason, but this sentence was commuted to life in prison. Petkov quotes Stamboliiski as saying that he was pressured initially to sign a declaration that he approved the foreign policy course and military action in order to preserve his life, but with the latter's refusal, incarceration was preferred in order to avoid revolts in the army.[18] Stamboliiski spent the war years in prison, and, although he had been relegated to strict confinement for life, the conditions of political prisoners in the Central Prison of Sofia allowed a softening of the blow against the BANU. Kosta Todorov was also incarcerated with Stamboliiski and he relates how the cells were unlocked and prisoners could walk around freely and visit each other. In addition, they read the Bulgarian and French press and were allowed "dentist visits" in town twice a week.[19] Only in 1918, when Stamboliiski's efforts to prepare agrarians in the army for a possible coup d'état was discovered did Radoslavov decide "to break up the Agrarian 'General Staff' in the Central Prison."[20] This treatment very much resembles that of Raskolnikov, Trotsky, Kamenev, and Lunacharsky in the Crosses Prison after the July Days in Petrograd.[21] Until then, Stamboliiski was actively writing and was engaged with providing direction and managing disputes within the BANU.

In the brochure "Power, Anarchy, and Democracy" that he wrote in prison in 1917, Stamboliiski prepared the ground for the agrarians to take the reins of government. Written seemingly as an abstract theoretical treatise, it was in fact a manifesto: "The democratic regime will call to the stage those hidden and until now sleeping national forces, will bathe them in the light of general enlightenment from school education, will temper them in the impetus of struggle, and will utilize them in the whirlpool, the swarm of creative political life.... Power wielded by the people can undo the evil that has been caused by power wielded by monarchs and oligarchies."[22]

Stamboliiski's intervention prevented the fractioning of the BANU's parliamentary group between the incarcerated leadership and those that had remained at liberty, and the united front allowed the resumption in the publication of *Zemledelsko Zname* from August 1917, although under strict censorship. When the Radoslavov government fell from the added weight of reversals in the war, the activities of the Opposition Bloc, which the BANU had joined, and the failure to recover Dobrudzha, the BANU did not join the Malinov cabinet when it was formed on June 21, 1918. Indeed, although the Opposition Bloc parties gave their support to the new government, Stamboliiski used the platform of *Zemledelsko Zname* to critique his colleague Dragiev's call for BANU to join a government of national unity. He was banking on his conviction that Malinov would not last and that the BANU would be able to take power, and so all efforts went into agitating soldiers at the front and making organizational preparations to rule for when the war ended.

The collapse of the Bulgarian front at Dobro Pole on September 15, 1918, set the stage for Stamboliiski's release from prison and his second meeting with Tsar Ferdinand. The Bulgarian soldiers mutinied on their retreat to Sofia and only the fact that Bulgarian forces were able to hold their fortified positions at Dojran on September 18 and 19 prevented occupation by the advancing Greek and British forces. The meeting of Stamboliiski and the king on September 25, 1918, completes the saga that began with the first audience in 1915. After exchanging a few initial "pleasantries" in which the king accused Stamboliiski and the BANU of stabbing the country in the back and destroying the front, while Stamboliiski reiterated that the blame for the national tragedy lay with the throne, the personal attacks mounted when the question of peace at any cost was broached. I am providing the content of some of the repartee in greater detail because it is so characteristic of the two mentalities of governance as well as the theatrical élan with which the changing of the guard is presented.

> [Ferdinand] I have been and will remain a true friend and ally of the Central Powers. In my veins flows the blood of the nobleman.
> [Stamboliiski] I do not doubt that, because I have tasted plenty of your nobility.
> [Ferdinand] You have the coarseness of the peasant! . . .
> [Stamboliiski] But I do not have the wickedness of the nobleman.
> [Ferdinand] You are unimaginably bold! . . .
> [Stamboliiski] But extraordinarily truthful.
> [Ferdinand] You our tongue is intolerable! . . .
> [Stamboliiski] But my heart and soul are captivating. I have, as the Russians say, a nasty tongue, but a golden heart, and you exactly the opposite . . .
> [Ferdinand] You make use of the most boorish threats against me!
> [Stamboliiski] And still, not one hair from your head has fallen because of me.
> [Ferdinand] You lack flexibility and perspicacity. You still have not reformed yourself . . .

> [**Stamboliiski**] It is too late, Your Highness, to show regret [*pishmanluk*]—I am out of prison. Besides, I find that your evaluation of me is incorrect. I feel that I am in the prime of my powers. I tested and tempered my will in the fires of the harshest suffering, and as to my intellectual prowess and political discernment, the events from 1912 up to now most eloquently attest to their scope and brilliance. During all those tumultuous and fateful times, that which I foretold, has happened. I also studied Your statesmen and councillors and I find, when I do an impartial comparison, way too many advantages in my favor. You will not find vacillation, uncertainty, or instability in my soul and character—something that acquires a most tragic shape in them. That you have reached the same conclusion is shown by the fact that the day before yesterday you sent a man to me with the offer to hand the government to me in these most fateful minutes for Bulgaria.[23]

When the conversation turned to what was to be done, it revolved around the issues of the reconstruction of the cabinet and the buttressing of the front until a ceasefire could be arranged. Stamboliiski positioned himself against the immediate forming of a new government, not because he was, as a matter of principle, "against the reconstruction of the present cabinet, but because I am convinced that it won't occur in the way that the needs of the times demand. For a real change in the cabinet to occur, a change that the whole nation wants, events have to somehow put greater pressure on you [Ferdinand]."[24] The turning point of the meeting was the proposal to pacify the troops at the front. Ferdinand was ecstatic and could not wait to have Stamboliiski depart. Only as a farewell courtesy did he ask what Stamboliiski intended to do after that. The response anticipated the rebirth of the BANU after the war: "My first task after the signing of the peace is to revive, cure, and recreate the Agrarian Union, after which I will apply all my energy to the deep national [*vsenarodno*] revival of unfortunate Bulgaria that experienced such long and unfortunate wars."[25]

The very next day, Stamboliiski and Raiko Daskalov, who had been imprisoned with him, left for the front in the company of the Minister of War and a few other parliamentarians. They met the returning rebellious troops in the village of Radomir outside of Sofia. While Stamboliiski continued south in order to reach the army headquarters in Kiustendil, Daskalov remained in Radomir. By the next day, Daskalov had organized the soldiers into eight infantry battalions and two machine gun companies, had proclaimed a republic, declared a provisional government, and had sent an ultimatum for recognition to Sofia. The rebelling soldiers slowly marched on Sofia but were routed on September 30 by Macedonian brigades and German reinforcements. Even with this defeat, the transformation of Bulgarian politics was not halted, as Tsar Ferdinand's forced abdication on October 3, 1918, in favor of his son Boris III pacified the rebellious atmosphere among the soldiers.

The Radomir Rebellion or the Soldiers' Uprising, as this episode has entered the historiography, generated contentious historiographical debates almost from its conclusion.[26] The central issue of the debate was whether the agrarians planned the proclamation of the republic or whether they were swept up in a popular uprising. Naturally, this was most important for communist analyses, and, indeed, Daskalov added Stamboliiski's name to the declaration with neither his knowledge nor his presence. Upon his return from Kiustendil, Stamboliiski did not join Daskalov, but rather returned to Sofia, where he had to contend with the refusal of the Agrarian Parliamentary Group, headed by Dimitur Dragiev, to recognize the rebellion.

Stamboliiski inflated his role in the rebellion for political purposes when he began speaking about it in 1919.[27] Raiko Daskalov exaggerated the historical significance of the episode in a speech on June 18, 1919, in which he talked about the "September Revolution" (not to be confused with the September Uprising of 1923).[28] In a skewed attempt at comparative analysis, which is, however, indicative of the connections and inspirations that the interwar agrarian movements derived from each other, he announced, "The Czechs are also a Slavic people like us, they too led a war against the Central Powers, and not for three years like us, but for four. Today, however, they are incomparably better off than we are, chiefly because they managed to break with the politics of the Habsburgs after the rout and follow up on the consequences. We embarked on the same road a month earlier. They succeeded, we didn't."[29] Kosta Todorov also took liberties with causality when he wrote, "Daskalov's rebellion had failed, but its main purpose was achieved. On the very day it was crushed, Malinov sent a delegation to Salonika to ask for an armistice, and persuaded King Ferdinand to abdicate."[30]

Much more important for the history of Bulgaria, however, were the tensions with Dimitur Blagoev's *Bulgarska rabotnicheska sotsialdemokraticheska partiia (tesni sotsialisti)* (Bulgarian Workers' Social Democratic Party—Narrow Socialists), which was to rename itself in 1919 as the *Bulgarska komunisticheska partiia (tesni sotsialisti)* (Bulgarian Communist Party—Narrow Socialists, BKP, t.s.), that emerged from this episode. Stamboliiski sought Blagoev's help to create an internal uprising in the capital during the days of the Soldiers' Uprising. He was prepared to accept the program of the communists apart from the assault on private property, but Blagoev's calculations were that Stamboliiski would occupy the role of a Kerensky and refused support.[31] This was the first step that turned the two parties which mobilized the revolutionary sentiment of the postwar years from potential allies into adversaries.

In terms of organizational transformation, the end of the First World War marked the resurrection of the BANU. During the war years, from 1915 to 1919, the union had not collected membership dues. The Permanent Presence (*Postoiannoto Prisustvie*) of the party, its executive body, did not function, and, alongside Stamboliiski, two of its other members, Stancho Momchev and Andrei Sharenkov, were incarcerated. Conscription, too, meant that the rank and file of the local agrarian organizations, the *druzhbi* (unions), spent the duration of the three wars at the fronts. It was in the aftermath of the armistice and demobilization, and the release of the BANU leadership, that the organizational life of the party could resume. When the 77,298 members of BANU in 1919 elevated it to the rank of the most numerous party in the country, one needs to take stock of the furious organizational activity: the agrarian *druzhbi* numbered 1883 at the start of June 1919, but nearly half of those, 920, had been formed after the war.[32] The amnesty at the end of 1918, whose promulgation was sped up by BANU and communist grassroots agitation, also opened the gates to the electoral successes of the parties of change.

BANU participated in the short transitional coalition governments of Aleksandur Malinov and Teodor Teodorov, and Stamboliiski signed the Peace Treaty in Paris. From January 1919, Stamboliiski was a coalition partner to the government of Teodor Teodorov. In his capacity as Minister of Public Domains, he had first been a part of the Bulgarian delegation to the Peace Conference in Paris. After the electoral successes of the BANU in August 1919, the BANU led the coalition which continued to include the National Party of Teodorov. Teodorov remained the head of the delegation, but he refused to sign the treaty and resigned in protest. In his capacity as prime minister, it was Stamboliiski that then signed the treaty. In the *Narodno Subranie* (National Assembly) elections of August 17, 1919, the BANU received the largest percentage of votes and 36 percent of the elected deputies. With the Communists in second place and the Broad Socialists in third place, these three parties occupied 71 percent of the seats in the *Narodno Subranie* after having received a combined 59 percent of the vote. Stamboliiski had to create a coalition government, and again he first approached Blagoev but was rejected as representative of the petite bourgeoisie. A coalition with the Broad Socialists was attempted next, and in Paris Stamboliiski reached an agreement with Ianko Sakuzov to create a cabinet of six Agrarians, three Broad Socialists, and one Radical. On their return to Sofia, however, this agreement collapsed because of the demands of the Broad

Socialists for four ministerial posts that would include the Ministries of Interior, War, and Foreign Affairs. Presented by the Broad Socialists ostensibly as a way to secure food deliveries and to combat speculation, this demand was taken as a thinly veiled preparation for a coup.[33] In this way, Stamboliiski was forced to work with the National and Progressive Liberal parties.

On the day that the new parliament was to begin work, December 24, 1919, the Communist Party planned demonstrations to challenge the government. As a result of the imposition of martial law, few of the planned demonstrations occurred, but the next day the railroad workers went on an unplanned strike. This in turn expanded into a general strike on December 28, 1919, and it received the support of the Broad Socialists. Whereas the strike grew from below and expressed genuine labor unrest, the BANU and the rest of the political spectrum responded to it as a political challenge. Using the Orange Guard, which was the agrarian militia, and Allied support, Stamboliiski decisively and mercilessly crushed the strike on January 5, 1920. The rupture between the two streams of the Bulgarian Left was complete and later found further expression in the communist refusal to defend the BANU regime against the June coup of 1923, despite the criticism from Moscow, something elaborated in greater detail in Chapter 4.

After the communist challenge, Stamboliiski dissolved parliament and called for new elections on March 28, 1920. This resulted in the BANU getting more than twice the deputies of the Communist Party, and, in fact, BANU was only five seats short of an absolute majority. Using a technicality in the electoral law, thirteen seats were invalidated, and, from May 20, 1920, the BANU embarked on its project to transform Bulgaria.

As a loser in the wars from 1912 to 1918, the Bulgarian case reveals the greatest radicalization and presents the traditional political structure at its weakest. Consequently, the transformative effects of the war almost enable a teleology that places the BANU at the head of the country in 1920. However, the cases of national triumph—rather than tragedy, like the ones in Yugoslavia and Czechoslovakia—were also subjected to the transformative pressures brought about by the war, especially when the questions of state formation and reorganization are concerned.

## Agrarians in the service of the nation and the republic

Even though it was founded after the First World War, the *Savez Zemljoradnika's* appearance on the political stage becomes intelligible only in relation to its challenge to and encroachment on the base of the NRS of Nikola Pašić. The NRS's roots in the Serbian village can be traced to the legacy of Peasant Radicals that was formed in 1874 as an oppositional group in the *Skupština*. The uniting of agrarian delegates around Adam Bogosavljević was an expression of peasant discontent aimed at addressing peasant grievances and the dismantling of the bureaucratic state. The Radical Socialist movement under Svetozar Marković and Nikola Pašić was sympathetic to Bogosavljević and joined the Peasant Radicals in the *Skupština* even before it had formed the NRS. But with the death of Marković in 1875 and Bogosavljević in 1880, Pašić was able to co-opt this foundation when he formed the NRS on January 8, 1881. Augusta Dimou summarizes this process: "By absorbing and appropriating the legacy of the Bogosavljević group, the Radicals managed to effectively stifle the rise of a genuine, grassroots peasant party. It is significant to underline that the populist socialism of the Radicals met the discontent and the collectivist aspirations of the agrarian world halfway, adopting therefore several of their concrete social and political demands, and echoing to a large extent their worldview. Along with other factors, this goes far to explain the large-scale success of the Radical Party."[34] Dimou is absolutely correct in characterizing the NRS as a populist peasantist formation that presented itself as the mouthpiece of the peasantry, and her analysis is that much stronger for recognizing that "both radicalism and agrarianism [here the comparison is to the BANU] shared an anti-capitalistic core credo and envisioned the retention of the small agrarian producer."[35] This aspect of the NRS ensured its hegemony, especially as it could also draw on the victory in the First World War and the enormous territorial expansion of the state.

That the *Savez Zemljoradnika* could challenge and carve out its own electorate in the Serbian village against such an opponent is a testament to the resuscitation of agrarianist sentiment immediately after the First World War. Serbia had lost 25 percent of its population in the wars from 1912 to 1918 and agricultural production fell by 70 percent.[36] Foremost on the mind of demobilized peasants was the question of land reform. In Bosnia and Hercegovina, one of the few areas on the Balkans that were characterized by large landholdings and feudal relations, there were peasant revolts and seizures of land in November and December 1918. In this context, one of the primary questions for the first government formed under the leadership of Stojan Protić on December 20, 1918, through a compromise between the National Council of Serbs, Croats, and Slovenes, and the Serbian government was land reform. Prince Regent Alexander issued a manifesto on January 6, 1919, that promised agrarian reform, the elimination of feudal relations, and the parcelization of large estates, and asked the populace for patience while the reform was achieved through legal means. The Preliminary Regulation from February 25, 1919, began the land reform.

As in Bulgaria, the question among the radicalized and politicized peasantry was one of political representation and the lack of faith that lawyers and merchants could represent them in parliament. On October 12, 1919, Mihailo Avramović, who had cut his teeth in the creation of the cooperative movement of the *Raiffeisen*[37] type before the war, along with Milan Vlajinac and Mihailo Popović, found the *Savez Zemljoradnika*.[38] Having settled on rules and a program, the *Savez Zemljoradnika* began publishing its newspaper, *Selo*, in December 1919. In Bosnia and Hercegovina, the *Težačka Organizacija* (Peasant Organization) was founded on August 25, 1919, in Sarajevo at a gathering of 154 delegates which were delegated from peasant assemblies in 43 districts. Its goal was radical land reform, separate from the national and religious frameworks. At its meeting on June 6, 1920, the *Težačka Organizacija* reorganized itself as a party under the name *Savez Težaka u Bosni i Hercegovini* (Union of Peasants in Bosnia and Hercegovina). However, at its congress in Sarajevo on December 5, 1920, the party decided to merge with the *Savez Zemljoradnika*. In fact, at the Constitutional Assembly elections of 1920, the *Savez Zemljoradnika* incorporated not only the Bosnian party but also the *Težački Savez* from Dalmatia, and the *Seljački Savez* from Croatia. Only the Slovenian *Samostalna Kmetijska Stranka* (Independent Agrarian Party) had its name next to the *Savez Zemljoradnika* but with the proviso that its delegates would work with the latter in the *Skupština*. After the electoral results, the *Savez Zemljoradnika* came in fifth place with 39 deputies out of the 419. The boycott by the CPPP meant that it actually was the fourth largest block.

In Croatia, the CPPP was the direct beneficiary of the radicalization of the peasantry during the war. This radicalization, which was the result of military casualties and military requisitioning in the village, meant that by 1918 the countryside was in revolt. Between October and December 1918, armed bands of peasants and returning soldiers, the *Zeleni Kadar* (Green Cadre) sacked estates and generally made the Croatian countryside ungovernable. This spontaneous action underscored the divide between traditional Croatian politics as advanced by the National Council and the taste of freedom from administration that the collapse of the Dual Monarchy produced. The National Council's policy to preserve order until the achievement of unification was thus at the center of the rural/urban divide. The CPPP, with its anti-unitarist program and republicanism, became the only organized political force that could address these demands. In addition, because the elections for a Constitutional Assembly only took place at the end of 1920, for two years the Croatian lands were administered through the Serb bureaucratic system. The transferred resentments could then latch on to the unique ideological position of the CPPP and its organizational strength at the local level in order to resist the intrusion of the state. Biondich is correct in pointing out that "the party's grassroots organizational work before 1914 paid dividends in the postwar era, but neither the strength of the CPPP's organizational machinery nor the importance of local party functionaries should be exaggerated."[39] For example, the party had not held its main assemblies at all during the war years. Still, a movement that Biondich estimates to have had 15,000 members in 1914, had 2000 local party organizations and over a million members by 1921.[40]

Reflecting on this state of affairs, Stjepan Radić wrote, "This war created a completely new opinion and conviction in our peasant people; the opinion that the root of all evil and in particular of this war are the kings [*vladari*] and the conviction that the entire government and administration must be changed in their foundations according to peasant need and peasant right."[41] The oppositional nature of the CPPP as the embodiment of peasant discontent is reflected also in the changing of its name in December 1920 to the *Hrvatska Republikanska Seljačka Stranka* (Croat Republican Peasant Party, CRPP). Its antagonism to Serb centralism also resulted in repression. Radić was imprisoned from March 25, 1919, to February 27, 1920, and then again from March 23, 1920, to November 28, 1920. He was amnestied on November 28, 1920, the day of the elections to the Constituent Assembly, in time to witness the emergence of the CPPP as the only significant party of Croatia. In the elections, the CPPP came third nationally in the number of votes and fourth in the acquisition of deputy seats (14.3 percent of the vote and fifty seats).

\* \* \*

In one sense, the role of the RSČV in the early years of the Czechoslovak First Republic can be seen as anti-revolutionary. It was a guarantor of order in the revolutionary years between 1918 and 1920. It was the strongest party in the Revolutionary National Assembly, but that position was based on the proportional application of the 1911 Reichsrat election results. At the first corrective in the parliamentary elections of April 18, 1920, the RSČV lost half its seats and was able to get only 9.7 percent of the vote. The merger with the *Slovenská národná a rolnícka strana* in 1922, when it changed its name to the *Republikánská strana zemědělského a malorolnického lidu* (Republican Party of Agriculturalists and Small Farmers), added twelve seats to its twenty-eight, but they were still overshadowed by the seventy-four seats of the Social Democrats. The Republican Party would only regain its primacy in the parliamentary elections of 1925. Nonetheless, I argue that its central role in establishing and securing the stability of the political system was revolutionary in its own right.

The immediate postwar years were marked by civil unrest due to shortages and inflation, yet, Antonín Švehla, the leader of the Republican Party, was able to use the party's position to maintain moderate parliamentary coalitions. Within his own party, he charted a middle course between wealthy landowners and poor peasants. To give an example, during the Karel Kramář government from November 14, 1918, to July 8, 1919, Švehla occupied the posts of vice premier and Minister of the Interior. As interior minister, he reduced urban unrest through the increase in the number of gendarmes and the formation of a "preventive service."[42] Miller contrasts the political paralysis during the Habsburg Empire to the moderate policy that was maintained in the difficult postwar years: "The catalyst for interparty cooperation and good relations among party leaders was Švehla. His political style had much to do with his success at negotiation. Even when he became de facto prime minister, he did not try to dictate policy but rather sought the support of other coalition parties in the form of a narrow ministerial committee, which brought together Agrarians, Social-Democrats, Czech Socialists, and State-Rights Democrats."[43]

Švehla's indispensability, and by extension that of the Republican Party, can be seen in his participation in the Red-Green coalition after the Kramář government fell. From July 8, 1919, to the elections of April 18, 1919, the Republican Party was a coalition partner to Vlastimil Tusar's Social Democratic Party based on its postwar strength in the Assembly. But even after the dismal results, Tusar's new government insisted in keeping Švehla in the coalition. Finally, when the growing communist wing in the Social Democratic Party threatened the stability of the government in the summer of 1919, Masaryk, Švehla, and Tusar decided to resolve the crisis through the appointment of a government of experts headed by Jan Černý. To support the Černý government, and ensure stability, Švehla instituted the Pětka that Daniel Miller described above. Quoting Ferdinand Peroutka, Miller characterizes it as, "the Pětka governed and the cabinet of technicians administered."[44] While the trajectory of the Republican Party did not capitalize on the social unrest following the war, its shift from reforming the monarchy and working through the Reichsrat for independence, and then for the maintenance of order, institution building, and the formation and defense of the political structure of the First Republic is at the heart of the extraordinary and unique perseverance of the Czechoslovak parliamentary order in the interwar period. During that time, the Republican Party participated in every coalition government.

## Prewar origins

Another, perhaps more evocative, way to express the transformed position that the agrarian movements occupied after the First World War is through a juxtaposition with their origins prior to the conflict. Except for the *Savez Zemljoradnika*, which was a new formation that could capitalize on the space opened by the war, the other movements had their beginnings in work that was quite limited in its political scope and whose initial impetus was provided by socioeconomic grievances. While it would be a mistake to discount the significant gains and growth that these movements achieved in the first years of the twentieth century, nonetheless the developmental trajectories of the agrarian movements after the First World War were so accelerated and transformed that one must treat them as changes in kind rather than degree. In other words, the catalytic effect of war becomes self-evident through the simple exercise of presenting the origins of the agrarian movements prior to it.

The BANU was founded on December 28, 1899, from a collection of peasant professional and cultural associations. The lack of agricultural credit and the rampant usury in the countryside had provided the impetus for the creation of these nonpolitical associations. Consequently, the BANU was organized initially not as a political party. Its charter called for the enlightenment of the peasantry through education and the establishment of mutual savings banks and the provision of cheap credit, in other words the intellectual and material improvement of the condition of the peasantry. Only at its Third Congress in 1901 did the party become political, but at the cost of the departure of some of the membership that was led by one of the founding fathers, Tsanko Tserkovksi. From the time the party's first program was adopted in 1905 (prepared by Stamboliiski), the BANU began to establish itself in political life, especially through its anti-monarchist stance and socioeconomic program. Nonetheless, at the parliamentary elections of May 25, 1905, it could muster only 11 percent of the vote, and even though after the Second Balkan War its parliamentary gains had doubled in the November 24, 1913, election, it still could only claim 21 percent of the vote.

The Republican Party, known then as the *Českoslovanská strana agrární* (Czech-Slavic Agrarian Party), was founded in 1905 from the fusion of the *Česká strana agrární* (Czech Agrarian Party) and the *Česká strana agrární pro Moravu a Slezsko* (Czech Agrarian Party for Moravia and Silesia). It worked to improve the condition of smallholders and one of its first successes, one that propelled Švehla to the forefront, was in the support it afforded to sugar beet farmers. The refining industry was dominated by wealthy estate owners, while the raw material was grown by smallholders. The refiners ran virtual monopolies that generated usurious abuse. Švehla worked within the Central Union of Sugar Beet Growers and in 1905 caused the deliberation of the problem in the Reichsrat. In the first years of the twentieth century, the *Českoslovanská strana agrární* created economic interest groups among other sectors of agriculture and that activity generated a popularity and base in the countryside. Although the electoral system expanded to universal manhood suffrage in 1907 in the Dual Monarchy, and that in turn began transforming the agrarians into a mass party, its economic activity was still central. The expansion of party membership from around 33,000 at the end of 1906, to 40,000 in 1908, to 91,194 and 2,500 local organizations in 1914 shows a party that was able to capitalize on the transformative political potentials that the consequences of the First World War brought.[45]

The CPPP was founded on March 10, 1904, by the Radić brothers because they were convinced that the peasantry needed its own expression, against the politics of the intelligentsia, to advance the cause of the Croatian nation. This was the reason that, unlike the other Croatian parties before the war, its base was organized in the village. From the adoption of its party program in 1905, the party sought to advance the social and political level of the peasantry while seeking a solution to the national question. The latter meant resistance to the unitarist projects with Serbia, and this position kept it on the political margins until after the First World War. For instance, its representation in the *Sabor* which had eighty-eight seats was zero in 1906, two in 1908, nine in 1910, eight in 1911, and three in 1913.[46]

By June 28, 1921, when the Vidovdan constitution was approved in the Triunine Kingdom, the principal agrarian players in Bulgaria, Czechoslovakia, and the Triunine Kingdom had already forged or recast themselves into the political movements that would occupy center stage in the interwar period. Already from May 21, 1920, the BANU had been able to establish majoritarian rule and embark on the ambitious reform programs that, as quoted earlier, allowed Stamboliiski to sum up the significance of his politics: "Today there are only two interesting social experiments: the experiment of Lenin and my own."[47] Similarly, Antonín Švehla had placed his stamp on the politics of the First Republic with the creation of the institution of the Pětka in September 1920.

Out of the ashes of the First World War, therefore, four autonomous mass movements had emerged that in a syncretic way would present the three faces of the alternative modernity imbedded within the golden age of the European peasantry. The Bulgarian case was the clearest expression of agrarian radicalism. The case of the Kingdom of Serbs, Croats, and Slovenes was dominated by the national question, the unifying efforts of the *Savez Zemljoradnika* notwithstanding. The Czechoslovak case, on the other hand, presents a moderate and centrist agrarianism that was the key guarantor of parliamentary stability. That is not to say that nationalism was not a factor in Czechoslovakia, and, indeed, the Republican Party could not properly address the Sudeten German problem in the way that it incorporated the Slovak agrarians. The key difference in the development of Czechoslovakia and Yugoslavia, which otherwise shared a similar task of nation building, lies in the fact that Serbia had had the experience and history as an independent state from the nineteenth century. In Serbia, the NRS had already challenged liberal politics before the First World War and to a degree had mobilized the peasantry. The victory in the conflicts from 1912 to 1918 thus provided it with sufficient prestige to hold out against the internal challenge of the *Savez Zemljoradnika* and the external challenge of the CRPP and this set the tone for the conflictual politics in interwar Yugoslavia. Conversely, Czechoslovakia was constituted after the First World War and, despite the greater development in the lands of Bohemia and Moravia, Slovakia was not subjected to the center-periphery politics that were in play in Yugoslavia. The unification of Czechs and Slovaks is metaphorically reflected in the renaming of the RSČV on June 29, 1922, into the *Republikánská strana zemědělského a malorolnického lidu* when it merged with the *Slovenská národná a rolnícka strana*. The completely new formation of the Czechoslovak state gave the space for the Republican Party to become the major keystone of the interwar system.

This chapter has located the formation of powerful agrarian political organizations within the context of the First World War and the disruptions it caused. It has argued that this context allowed the agrarian parties to project themselves onto the national stage, often as the most powerful political entities, and that this newfound significance quantitatively and qualitatively transformed these organizations. The next chapter extends this argument to the international stage and looks at the competing interests and initiatives behind the projection of the peasant on the world stage through the MAB. In both cases, the merger of an articulated peasant subjectivity with organized political action is the conditio sine qua non for this surge.

# 2

# Projecting the Peasant on the World Stage

On April 10, 1921, the *New York Times* ran an article entitled "The Little Anti-Red Entente." The subtitle, "Protective Alliance Engineered by Jonescu and Benes—'Green Internationale' Plan," added another layer to the hodge-podge coverage of developments in Central and Southeastern Europe as they appeared from New York. As a result, this article was memorable not only because it was one of the first treatments of the idea of a Peasant Green International in the press but also for the way it rather clumsily tried to link it to the Little Entente through the idea of an ascendant anti-Bolshevik front. The article eschewed the representation of the Little Entente, which was signed on August 14, 1920, as the defensive alliance between Czechoslovakia and the Kingdom of Serbs, Croats, and Slovenes that Beneš envisioned against a revanchist Hungary and the possibility of a restoration of Austria-Hungary. Nor did it spill any ink on the refusal of Romania to join the alliance on August 17, 1920. Romania in fact became a signatory only on April 23, 1921, almost two weeks after the article appeared. While France supported the Little Entente, its role was promoted to being decisive by the article's author, Walter Littlefield:

> The "Little Entente" is only one of several more or less binding or moral agreements between the emancipated States of the old Dual Monarchy and the neighboring States which have chiefly profited by its partition. . . . Often working at cross-purposes but generally with identical objects in view, these agreements have been mainly directed by France and Italy, who have more or less shaped the initial programs of Take Jonescu, the Rumanian statesman, and Dr. Edouard Benès, the Foreign Minister of Czechoslovakia, and the *interesting scheme of Alexander Stambulisky, the peasant Premier of Bulgaria, for a "Green Internationale"—a farmers' alliance which should oppose the "White Internationale" or reactionism, on the one hand, and the "Red Internationale"* of *anarchy and Bolshevism, on the other.* (italics mine)[1]

A consequence of the thesis that the West was constructing and directing an anti-Bolshevik front was that the interests and agencies of the states of Central and Southeastern Europe were being written out. Furthermore, the imperative of imagining a bulwark against the Soviet state dictated the streamlining of convenient facts and the subsumption of contradictions. While the Little Entente falls prey to this constricting logic, some redemption could be found in the context of Bela Kun's Soviet republic in Hungary. The liberties taken with the insertion of Alexander Stamboliiski's regime and the Green International were a different matter altogether.

Before performing the critique of this representation, I would like to present in a little more detail the specific "contextualization" that the article offers vis-à-vis Bulgaria. It is a useful departure point because the conventions of reporting in the press are the condition for the creation of a text that imbeds within it the keys to its deconstruction.

In line with the stereotype of Bulgaria as the bête noire of the Balkans, the article reaffirmed the image even as it suggested that the scheme for the Green International was a sign that finally the country was moving in the right direction. "The position of Bulgaria—her correct attitude in trying to carry out the terms of the Treaty of Neuilly and her subsequent hope of more territory under the readjustment of the Treaty of Sèvres, her supposed leanings toward a sort of rural Bolshevism has, meanwhile, been measurably elucidated by the recent visit of M. Stambulisky to Prague, Warsaw and Bucharest, and the formulating of his scheme for a great 'Green Internationale,' support to which was pledged him in those capitals."[2] The description of the theory of the international, as the article presented it, was based on a speech by Stamboliiski at the "Congress of the Agricultural League" on February 15, 1921, which the article paraphrases as follows: "This was to be an international union of the peasants of Central and Southeastern Europe, to offset and overcome the work of the White Internationale of the reactionaries, who wanted to restore the dethroned monarchs, and also of the Bolsheviki (the Red Internationale), who were attempting to destroy all government and with it industry and individual initiative in industry as well as in the arts."[3]

The quotation from Stamboliiski that was provided was decontextualized and offered in a way to suggest that the Green International is a direct response to the threats of the "reactionaries" and "Bolsheviki" in the passage above: "I have no doubt that our 'Green Internationale' will ultimately free Russia from the Soviets. At least, it is destined to free farmers elsewhere from the unjust restrictions placed upon them by the manufacturers and capitalists, who know nothing about farming, and make both realize that the farmer is just as necessary to the life of a country as the workers on the roads, in the factories, or employed in transportation."[4]

It is fascinating how what originated as a direct critique of liberal capitalism was deflected toward the monarchists of the ancien régime. Even more incongruous was the manipulated "meaning" of freeing Russia from the Soviets. A Western reader might not be aware that Stamboliiski had begun an article entitled "The Russian Revolution" in *Zemledelsko Zname* in 1918 with a glowing assessment: "The most glorious event of these times remains the Russian Revolution. For now it provides a twofold service to the world: on one hand, it destroyed forever the most terrible tsarism, which endangered the progress of all humankind, on the other, it made the question of peace the order of the day."[5] These words cannot be reduced to temporary and misguided revolutionary euphoria carelessly tossed from a prison cell in 1918; one only needs to look at his speech on Bulgaria's foreign policy that he read to parliament on October 26, 1921, for a refutation of the anti-Bolshevism claims. "And I have said that we have absolutely nothing against Russia, because we do not want to meddle in its internal affairs—let her rule herself as she sees fit: whether it is Bolshevik rule or something else, let us only hope it is not tsarist rule . . . . Because this tsarism ruined it, and not Bolshevism—tsarism ruined it."[6] Responding to his critics, Stamboliiski continued,

> You accuse us of Bolshevism and these accusations go into the mouth of foreigners. Supposedly we have been pushing Bulgaria towards Bolshevism. This accusation, this suspicion still exists. Why? Because we chase away the *Wrangelisti* [these are the White Russian remnants of General Pyotr Wrangel's army that sought refuge in Bulgaria after its defeat in the Russian Civil War], because we chase those generals which want to conspire in Bulgaria, which try to impose on us another mentality, another constitution, other laws. What do you want? You want us to pass a law with which to forbid the existence of these people. We cannot impair the freedom of the Bulgarian people. We want there to be a proper Switzerland on the Balkan peninsula. You strove to make it a Prussia, a Japan, we will strive to make it a Switzerland.[7]

As an anti-Bolshevik initiative grotesquely married to the Little Entente, the representation of the Green International in the *New York Times* was an artifact of wishful thinking and the need to make the complex and radical developments in the Balkans conform to the mental frameworks of a Western readership. However, I do not intend to single out a "Western" misrepresentation that is implicated in imagining the Balkans and then facilely overturn it through recourse to a Rankean positivism that draws its claims of authenticity from being rooted in the region. Substantive references to the Green International are minimal prior to 1923. I would like to illustrate this condition of flux and indeterminacy with three examples from three of the most prominent architects of that institution, Antonín Švehla, Milan Hodža, and Alexander Stamboliiski. The provision of this caveat will go a long way to explain how the space for outside appropriations such as that of the *New York Times* became possible. It also makes the "imagination" of the *New York Times* just one among several imaginaries that were sparked by the agrarian potential.[8]

The lack of clarity about the international, but also the interest it sparked, was evidenced by a letter sent by the Bulgarian Legation in Paris to Sofia on August 11, 1921. The letter informed Stamboliiski, who was also holding at that time the portfolio of the Ministry of Foreign Affairs, that the French government had begun to be interested in the agrarian movement in Europe and was seeking to determine if it represented a danger to the current social order. Not only did the letter report that the French agrarian parties had been investigated for possible links to the Green International but also that a French commissar from the interior ministry had questioned the secretary of the Bulgarian Legation about the goals of the international and the means with which it intended to pursue them.[9]

Until the MAB started its work seriously in Prague in 1923, the idea of the Green International was nebulous, fuzzy, and in the service of competing interests. Yet, even that crystallization of the institution was problematized in the internal self-assessment that the MAB produced in 1927. At the end of 1927, Karel Mečíř, the specially appointed secretary general of the MAB by Antonín Švehla to overhaul the institution, wrote an official letter to Jovan Jovanović, the president of the *Savez Zemljoradnika*, to inform him of changes in the statutes of the organization. The task of reorganization and enlargement of the activities of the MAB that Mečíř undertook were justified in that letter through a rather ungenerous summation of the first few years of work: "You know, dear Minister, that the hopes we had placed on the activity of this international organization were not realized at its inception. It is true that even the *Bulletin* appeared rather irregularly. In these conditions, one should not be surprised to see that the links between the Bureau and the member parties have weakened more and more, until they have become a pure formality."[10]

The fear of Mečíř was that without an international organization to bind the peasant political parties, not only would the political influence of peasants be diminished but also any other agrarian organization would be condemned to stagnation. Mečíř's letter specifically named Švehla as the founder of the MAB and there was no mention of Stamboliiski. Further, Mečíř referred to Jovan Jovanović's predecessor, Avramović, only as a signatory to Švehla's institution.

In a letter dated ten days earlier, October 15, 1927, this time signed by both Mečíř and Švehla, the elision of Stamboliiski's contribution to the Green International was even more glaring. While making the case for the necessity of an international agrarian organization, both authors used the bloody coup against Stamboliiski and the even more violent White Terror in its aftermath to engage in some counterfactual speculation. "And let us also remember that in Bulgaria, under a past regime, thousands of peasants were killed without due process, that a war of extermination was declared on the peasant political organization, the Bulgarian Agrarian National Union, a war which took the lives of twenty thousand peasants. Who learned of it, who protested? How many thousands of these wretches could have been saved if the peasant parties had been informed, united internationally and had decided to act?"[11]

As an appeal to international peasant unity, these words are compelling. However, Švehla was fully aware of the extent and consequences of the coup against Stamboliiski, that Czechoslovakia limited its assistance to the acceptance of a few political refugees, and that it promptly recognized the new regime of Alexander Tsankov. I will go into greater detail about the response of Czechoslovakia and Yugoslavia to the coup in Chapter 4, but the decision to adopt a passive response that would not shake up the postwar order requires a mention here to contextualize Švehla and Mečíř's next sentence: "All these circumstances and considerations *convinced us to create the International Agrarian Bureau. The International Agrarian Bureau in Prague exists since 1923* and is responsible for the appearance of the *Bulletin of the International Agrarian Bureau* which is to date the only international organ of the peasant political movement [italics mine]."[12]

As an imaginary that frames the function of the MAB, this one is no less problematic than the earlier production by the *New York Times*. But let us now complete the picture with an even more evocative iteration that was generated in Bulgaria by Stamboliiski himself a few short months before his death. Mihail Genovski was an eminent cadre of the BANU in the second half of the twentieth century. He served as the Minister of Agriculture and State Property in the second cabinet of Kimon Georgiev after the Second World War and was a member of parliament from 1945 to 1949 and from 1981 to 1990. In 1923, he was twenty years old and as a founder of the Bulgarian Youth Agrarian Union, he was invited to attend a meeting with Stamboliiski on April 2, 1923.[13] The purpose of the meeting was to coordinate with the youth movement the changes planned in the makeup of the BANU leadership, to begin preparation for changes in the political and economic program of the party, and to prepare the ground for the elimination of the monarchy and a new constitution. Alongside these topics, however, the question of the Green International arose in a context in which the leading cadres of the youth movement were asking Stamboliiski for clarification for the otherwise nebulous pronouncements that had accumulated concerning the Green International in proclamations in the press and in political speeches. This source is a unique interior view into the expectations and strategies concerning that initiative from the Bulgarian side.

The discussion began from an examination of the differences between the various agrarian parties in Europe and the absence of a common ideology to link them. When the difference between the Bulgarian and Czech parties over the membership of large landowners was discussed, Dimitur Iliev asked, "Why are we then uniting with them and creating a peasant international, the Green International?" Stamboliiski's tripartite answer lays bare the imaginary of Bulgarian agrarian interests:

> For us, for now, the Green International has a threefold significance: first, we have to secure the support of the political currents in other countries, and especially in those countries which are in the system of the victors, for example those that have entered into the Little Entente; second, when we interact with them, we propagandize our ideas and we can influence the social composition and the ideas of these peasant parties; third, the Green International is necessary for us as a means to differentiate ourselves from the communists.[14]

Each of these points is pregnant with signification. First is the projection of agrarianism on the international stage as a means to modify the postwar Versailles divisions between the victors and the vanquished and to supplant them with a cooperative, agrarian structure. Second is the expanding horizon of possibility of the radical reorganization of social structures and the implementation of a peasant alternative modernity. Third is the carving out of space at the same level of prestige as the competing developmental paths, in this case, namely, the communist alternative because it was the closest competitor in recasting the role of the peasant citizen. Further, while key terms such as the "Little Entente" and "communists" reappear between the different imaginaries of the Green International, the differences in the approaches to them is staggering. At its extreme, rather than an anti-Bolshevik front, the

Green International is conceived here as a parallel developmental front. Earlier in the meeting Stamboliiski categorically repeated his position concerning the Communist Party: "We would be no genuine representatives of the sovereign people [*narodovlastnitsi*], we the agrarians, if we forbid the legal life of a political organization, which, according to our tenets about the estate organization of society, has the historical right to exist and has deep roots among the working masses."[15]

The history of the crystallization of the Green International as an institution in the fall of 1923 is explained by multiple currents in the development and implementation of agrarianism in the first decades after the First World War and the responses of various peripheral forces. Two frames are useful for contextualizing that history: the federalist idea itself and the competition from its rival, the Moscow-based IPC (*Krestintern*). Both frames operated on the international stage where the agrarian parties, via the Green International, sought to project themselves. But that international stage, despite the hopes and efforts of the agrarian parties to recast it (foremost among which were the efforts of Bulgaria), was severely constrained by the Versailles system.

At the Paris Peace Conference, Stamboliiski appealed to the Great Powers for a just peace, which would not produce a truncated Bulgaria burdened with impossible reparations. On November 22, 1919, Stamboliiski, who was present in Paris to sign the treaty, sent out several letters with that goal in mind. The letter to Georges Clemenceau was an assurance of the good faith of the Bulgarian regime particularly since Stamboliiski could mobilize his own and the BANU's opposition to the policies of Ferdinand. It was also an appeal for clemency in these circumstances, since, as Stamboliiski argued, a just peace would eliminate the points of contention between the Balkan nations: "The League of Nations could establish more equitable borders on the Balkan Peninsula, acceptable to all nations, and in this way it could contribute to a final understanding and political rapprochement between the Balkan states. The victors could also solve the Thracian question by leaving Bulgaria a free and sufficient territorial outlet to the Aegean, which is necessary for our trade relations with them."[16]

In order to strengthen his appeal, Stamboliiski also wrote to his Balkan colleagues, Nikola Pašić (Serbia), Constantin Coandă (Romania, stand-in for Ion Brătianu), and Eleftherios Venizelos (Greece). The letter to Pašić deserves a closer look because its tone is pregnant with heartfelt affection for Serbia and presumes that that the feeling was mutual. The other letters are measured and depend on careful arguments. This one, however, begins with, "I believe that a voice will not be raised in Serbia to challenge my sincere desire for a tight rapprochement between the two Slavic nations—the Serbian and Bulgarian. My whole past as a political figure is an unbroken chain of deeds towards the realization of this idea. I fought against this fatal war in parliament, in the press, among the people and I have always preached not only rapprochement, *but a union between our two nations* [italics mine]."[17] For Stamboliiski, the postwar elevation of the BANU to a position of power meant not only a recasting of the domestic structures in Bulgaria but also the chance to shape the Balkans and Europe. With this in mind, he concluded his letter to Pašić: "Give us your hand. I will do in my fatherland everything possible to achieve this ideal and to put forever an end to the bloody struggles between our brotherly nations. Give me your valuable cooperation. Nothing is impossible! As long as there is good will, the basis for a lasting agreement will be found. In it is the salvation of the Balkan nations."[18]

Pašić did not even dignify this gesture of cordiality with a reply. The less ebullient letter to Venizelos, however, at least was answered. Writing on November 25, 1919, Venizelos politely, but firmly, rebuffed the suggestion of concessions from the Greek side and stood by the consequences of employing the principle of "War Guilt."

> If some sacrifice is necessary, Bulgaria cannot ask this sacrifice from its neighbors in accordance with its friendship. On the contrary, she is expected to give guarantees so that the doubts of her neighbors can disappear, doubts which are justified by the fact that everyone suffered from her perfidy. . . . that, which Bulgaria can hope for, is that with the exact execution of the conditions of the Peace Treaty and with the possible reparation of the damages caused by her fault, and also through the passage of time, the bitter memory of the misfortunes she caused to her neighbors [will] be forgotten. This is the way the relations of Bulgaria with her neighbors will breathe a contagious confidence, which will gradually grow [*sic!*].[19]

The failure of Stamboliiski to secure support for a new postwar order in the Balkans, let alone advance the further step of a Balkan federation, had two consequences. He returned from Paris isolated and burdened with the suspicions of his neighbors and the Great Powers that Bulgaria would try to evade its treaty obligations. On the domestic front, his failure was exploited by the forces arrayed against him, and, in addition, it made the outlook for economic recovery of the country dismal, given its isolation and heavy reparations burden.[20] The need to postpone or renegotiate the reparation payments was one of the greatest challenges that faced the BANU regime. At the same time, after numerous failures, for example at the Genoa Conference, this issue produced one of the regime's greatest foreign policy successes. On March 21, 1923, the Bulgarian government succeeded in signing a protocol with the Reparations Commission that divided Bulgaria's obligations in two parts. The first part, composed of 550 million gold francs, was to be paid by Bulgaria at 5 percent interest for 60 years starting on October 1, 1923. The second part, consisting of the remaining 1.7 billion francs, was to be frozen until 1983.[21]

By agreeing to immediately start paying off Bulgaria's prewar debts, Stamboliiski had been able to delay the start of the reparation payments until June 1922. With his characteristic sympathy and wit, John Bell relates Stamboliiski's efforts at that time:

> Having encountered a total lack of sympathy from "his" Reparations Commission, Stamboliski was astonished when the commission in Austria ruled that that country was unable to pay. Seeking to fathom the mystery of Austria's good fortune, he sent Kosta Todorov and Alexander Obbov to Vienna to investigate. They found that the Austrians had provided the members of their commissions with large cash bribes, valuable art works, and other amenities. Somewhat surprised that this part of his "schooling in diplomacy" consisted of a lesson long familiar to Balkan politicians, Stamboliiski decided to test it at home. The commission looked in favor on these new arrangements and submitted a report that led to Bulgaria's receiving a further postponement to the spring of 1923.[22]

Kosta Todorov tells that in the scramble to delay the start of reparation payments that were to begin in June 1922 for the amount of 105 million gold francs (the budget for 1921 had been 150 million), he delivered a letter from Stamboliiski to Poincaré on May 11, 1922. The offer in the letter, which Poincaré rejected, was for Bulgaria to pay ten million gold francs a year for three years, to be followed by a reassessment based on Bulgaria's economic condition. This alternative means of "influencing" the Reparations Commission cost the Bulgarian government twenty-five million leva, one tenth of the amount offered to Poincaré.[23] Stamboliiski's "corruption" will receive special attention in Chapter 5.

The solution to the reparations question in 1923 was tied to the limited successes Stamboliiski made on the international front to break Bulgaria's isolation. In March 1923, Bulgaria announced that it could not pay despite the threat by France and Italy that in such a circumstance they would impose sanctions that would topple the BANU government. When Bulgaria's neighbors were authorized to collect these sanctions through the occupation of the coal mines in Pernik and the seizure of customs as per the articles of the Treaty of Neuilly, Yugoslavia declined in order not to ruin her improving relations with Bulgaria, and so did Greece and Romania.[24] The roots of the détente which made this possible lay in Stamboliiski's famous foreign relations blitz, the 100-day tour of Europe, during which the arrangements were made for the formation of the Green International.

Starting on October 1, 1920, Stamboliiski departed on a diplomatic tour which would take him from London to Paris, Brussels, Prague, Warsaw, and Bucharest. In Great Britain, he met with Lord Curzon, Churchill, and Lloyd-George. In his report to the Ministry of Foreign Affairs in Prague from November 1, 1920, the chargé d'affaires in Sofia, Künzl-Jizerský assessed the difficulties facing Stamboliiski: "In any case the tenacity of Stamboliiski has to be recognized, with which he strives to lead Bulgaria out of today's political isolation, and that despite the fact that his zeal is met with big hurdles. For now, the international position of Bulgaria remains quite unfavorable."[25] Before he reached Prague on December 12, 1920, Stamboliiski passed through Paris and Brussels. On October 29, 1920, Mečíř, who was serving as Czechoslovak ambassador in Brussels (from 1926 he was at the helm of the Green International-MAB) wrote to Beneš about his conversation with the Yugoslav ambassador there, Marković. Marković expressed concern about a possible breakthrough in Bulgarian-Czechoslovak relations: "In my opinion this is coming too early. I will tell you that it is not enough to wait years; my assessment is that a new generation needs to come, a new Serbian and a new Bulgarian generation, which could have faith in one another. We have suffered too much, to have faith."[26]

Given the relations and obligations toward Yugoslavia as its Little Entente partner, the Czechoslovak position necessitated assurances. On October 30, 1920, a circular telegram was sent from the Ministry of Foreign Affairs in Prague to the Czechoslovak embassies in London, Copenhagen, Washington, Paris, Rome, Bern, The Hague, Bucharest, Budapest, and Belgrade. It read:

> We want to support an agreement between Bulgaria and Yugoslavia, however in a manner that is completely correct and loyal with respect to the Kingdom of Serbs, Croats, and Slovenes, and the Rumanians. We stand by our views concerning the excessively harsh peace terms for Bulgaria. However, if we were forced to choose concerning the Bulgarian-Yugoslav question, we would walk with the Yugoslavs as our closest allies. This, by the way, is a tactical directive and changes nothing in our opinion. This opinion, however, must not be manifested in such a way that it could be used to agitate or provoke.[27]

Indeed, on December 21, 1920, Stamboliiski telegraphed from Prague that he had been told that the previous day the Yugoslav representative in Prague had protested in front of the ministry that not only the Czechoslovak people but also the government had received him most hospitably.[28]

During his eight days in Prague, Stamboliiski met with Masaryk, Prime Minister Jan Černý, Švehla, and numerous members of the Republican Party. While the documentation regarding the 100-day trip is one of the comparatively richest periods of Stamboliiski's foreign policy initiatives, it is nonetheless scant. His telegram from December 16, 1920, speaks of a two-and-a-half-hour conversation with Karel Kramář, the head of the National Democratic Party, and only adds, "He is won for the cause."[29] Even the detailed and exhaustive report of J. Šrom, who was attached to Stamboliiski by the Czechoslovak Foreign Ministry and appears to have attended all the major meetings of the Bulgarian premier, only states that in his meeting with Masaryk, Stamboliiski mentioned that even in these difficult circumstances, he still had no second thoughts about his willingness to enter into a Yugoslav-Bulgarian federation.[30] Šrom concludes his report with a characteristic of Stamboliiski:

> By and large, the prime minister Stamboliiski can be characterized as a person with strong will, who does not stop even before violence, so long as it leads according to him to the goal. A person with extremely practical purposes, clearly with the qualities of an economist, without particular interest about questions that do not have exclusively economic character. His philosophy is natural, consistent, and at the same time ruthless, [he] serves his idea with his whole being and therefore is ready to sacrifice everything for its realization, even his popularity and his life. As an orator he is sharp, temperamental, [and] with considerable confidence and preparedness.[31]

Despite the difficulties he faced, Stamboliiski did much to soften Bulgaria's isolation through his tour. It is important to note that at that time, when Bulgaria's isolation and Yugoslavia's suspicion made the pursuit of federation impossible, the advancement of the Green International picked up that idea on a different front. The Green International could thus serve as a way to unite the agrarian Slav nations and serve a springboard to overcome the immediate post–First World War hurdles to the federal project. Had Stamboliiski survived past 1923, the history of the federative idea could very well have looked much different from how it is presented below.

\* \* \*

Since the early modern period, ideas have floated about the creation of a cooperative political system in Europe.[32] In the nineteenth and twentieth centuries, the period of nationalism's high tide, there were several attempts to implement these ideas in practice. Most of them were attempts to foster cooperation between rising national movements against what was seen as the collaboration of the repressive empires in which they were subsumed. In Central and Southeastern Europe, Kossuth's Danubian Federation movement in the decade after the 1848–49 revolutions, explicitly spoke of federation, but was, in fact, doomed by the fact that Magyars, Romanians, and South Slavs were interested in the movement only insofar as it satisfied their national aspirations.[33] While the Habsburg Empire opted finally for dualism with the Ausgleich of 1867, and not for federalism, the federalist project remained alive in the ideas of the Austromarxists. The idea of cooperation without the federative component was revived, although on a modest scale and with a predominantly anti-revanchist goal (preventing Hungary, Austria, and Germany from territorial revision), by the Little Entente between Czechoslovakia, Yugoslavia, and Romania in the 1920s and 1930s.

In Southeast Europe, the federative idea began with the attempts of the Balkan revolutionary movements to coordinate their efforts against Ottoman rule. The idea for mutual military help was closely linked to the establishments of future political unions and eventually of a Balkan federation. Inspired by the French revolution, the plan of Rhigas Velestinlis for a restored Hellenic Republic was arguably the first roadmap of these ideas. Throughout the nineteenth century, these belonged predominantly to the democratic wing of the revolutionary movements with their ideal of a constitutional republican democracy and had Switzerland and the United States as their model.[34]

As in Central Europe, there were several attempts to forge regional cooperation short of a federation, such as the First Balkan Alliance system in the 1860s and 1870s, and the Second Balkan Alliance system, which prepared the Balkan Wars. By the end of the nineteenth century, a Balkan federation was championed seriously only by the socialists, who were members of the Second International. Aside from the agrarians of Stamboliiski, they were also the only consistent force opposed to war. At the height of the Balkan War in 1912, Yanko Sakuzov, a socialist deputy in the Bulgarian parliament stated, "We do not want a Balkan Confederation created as a result of the war. What we want, what we are preparing is a Confederation uniting in fact all of the Balkan nations, including Turkey, for a work of peace, of labor, of production and exchange, a work of liberty and of progress.... Will you who are allied today, not turn against each other tomorrow as is already foretold in the press and diplomatic circles?"³⁵

Both the First Balkan Socialist Conference in Belgrade in October 1911 and the Second Balkan Socialist Conference in Bucharest in July 1915 passed manifestos calling for a Balkan federation. After the end of the First World War, the split within the socialist movement, existing already from 1903 between Narrows and Broads, was institutionalized with the creation of a socialist and a communist party. While the socialists spoke vaguely of a "rapprochement among the Balkan peoples and their union in a federation of independent States," the communists advocated for a "Balkan Socialist Republic" established through a proletarian revolution and the dictatorship of the proletariat.³⁶

The only other consistent anti-war and pro-federalist force was the agrarian alternative. In the face of snubs from his neighbors and the international community, Stamboliiski persisted in working for Balkan cooperation and unity. He specifically aimed to achieve improved relations with Yugoslavia and, if possible, even a federation.³⁷ He had allies in the shape of the Croatian Peasant Party, adamantly opposed to the centralism of Belgrade. In 1921, Noel Buxton, a British Labour politician, reported that "the party led by Radich—republican, Croatian and federalist—is rapidly growing in influence. Radich has proclaimed his desire to unite all the peasants of the Balkans in one state, and he especially invites the kindred branch of the Southern Slavs—'our Bulgarian brothers.'"³⁸

A curious episode of the interwar period was the alliance of a branch of the Macedonian IMRO (Internal Macedonian Revolutionary Organization) with the *Comintern*. The IMRO was instrumental in the overthrow of Stamboliiski because he had signed the Treaty of Niš and because of his acceptance of the status quo in Macedonia. Soon, however, part of the leadership realized that IMRO's aims of an autonomous Macedonia ran contrary to the stated goals of the Bulgarian government for annexation. The result was the negotiations with the *Comintern* that likewise supported Macedonian autonomy in the spring of 1924, which to this day are the subject of a bitter dispute. As of July 15, 1924, however, a bimonthly publication—*La fédération balkanique*—started appearing in Vienna in all Balkan languages, in addition to French and German. The program of the publication was defined in its editorial:

> The liberty and peace of the Balkans, through the Balkan federation, will only be attained by movements for national liberation which will break, as soon as possible, the bonds which attach them to the European and Balkan governments; which will hasten to unite, under their flag, the working masses of their nation

into a united national front; which will have aided and drawn upon their power for the social struggles of these same masses in neighboring countries; which, finally, will be eager to unite its forces into a single Balkan front directed against chauvinism and conquering imperialism from whatever quarter it may come.[39]

It was in the same publication a few months later that Stjepan Radić shared his vision on the Balkan federation:

> The Balkan federation can be only peasant and republican. It cannot have any trace of Rumanian or Hungarian feudalism any more than it can be a copy of Russian bolshevism. At the outset and probably for a long time it will not include Rumania or Greece because it is really and formally limited to the four principal Yugoslav peoples, the Slovenes, the Croats, the Serbs, and the Bulgarians, and to Macedonia and Montenegro, and finally to Albania, all this, naturally, with their complete consent.[40]

This vision proved to be short-lived. By the spring of 1925, fearing the destruction of his party over the smear campaign following his trip to Moscow, Radić committed a volte-face and recognized the monarchy, the dynasty, the Vidovdan constitution, and the unitary state.[41] That these ideas were not realized is beyond the point. As Stavrianos perceptively shows, "for the first time the Balkan federation movement had secured a mass basis, at least in Bulgaria and Yugoslavia. Hitherto Balkan cooperation and federation had been the dream of isolated idealists and of powerless revolutionists, or the slogan of diplomats and statesmen who almost invariably were interested primarily in national aggrandizement. For the communists and socialists and agrarians, however, federation constituted a fundamental and integral part of their program and philosophy."[42]

With the suppression of the genuine federalists—communists, socialists, and agrarians—the movement for cooperation lost its radical edge and assumed what several authors have called a liberal alternative.[43] The Third Balkan Alliance System started in 1930 with a series of several conferences aimed at furthering social, economic, and cultural cooperation between the Balkan states. Despite their success, there was practically no federalism in the conferences. But the attempt to conclude a Balkan Pact (the Balkan Entente) in 1934 turned the effort into an anti-revisionist bloc against Bulgaria. With the Great Depression, the economic penetration of Germany, and the coming war, these last attempts came to a halt.

I have argued that Stamboliiski's idea of a South Slav federation contextualizes the birth of the MAB through the hope that this idea could be pursued by other means. The postwar isolation of Bulgaria put an end to the Yugoslav federative idea of Stamboliiski as a realizable option in the near term. With the coup d'état in 1923, and the reversal of the Treaty of Niš, that option was gone. It survived only as a utopian vision, an alternative, even as a means for the BANU to differentiate itself from the policies of the Bulgarian monarchy as late as 1937.[44] The fall of the BANU regime removed this component from the imaginary of the Green International and over the course of its development under the leadership of the Republican Party, it never exceeded a level of cooperation that might either endanger Czechoslovakia's foreign policy or compromise the unitary state structure of the country. Thus, it is ironic that the dissolution of the MAB in 1938 (see Chapter 6) would also be marked by the renaissance of the federal idea by none other than one of the highest functionaries and ideologues of the MAB, Milan Hodža. One can see how the focus of foreign policy interests shifts from federation to international and back to federation over the course of two dozen years.

Milan Hodža published his *Federation in Central Europe: Reflections and Reminiscences* in 1942.[45] Hodža was careful to state in his preface, "My suggestion concerning a federation of independent countries of Central Europe is just a suggestion.... My suggestion is not made on behalf of any persons or any party, but on my own account."[46] But he concluded, "I also know that a spontaneous co-operation of all these countries is already strongly rooted in the minds and feelings of very many, so that it may emerge one day as the principle feature of Central European political aims to come."[47]

The first part of his book was a history of international cooperative efforts that begins with the efforts to reform Austria-Hungary in the early years of the twentieth century and then covers the interwar period through Czechoslovakia's domestic and international policies as a basis for cooperation. The second part of the book was an elaboration of the structures of his federation in Central Europe that would unite Austria, Bulgaria, Czechoslovakia, Greece, Hungary, Poland, Romania, and Yugoslavia. Hodža's proposal was structured around the tenets of agrarianism as it developed in Czechoslovakia—economic cooperation, the strengthening of the middle class through rural democracy, and so on. His section "From Struggle to Tolerance and Compromise to Co-Operation," was a direct extension of Švehla's politics of compromise in the First Republic.

Aware as Hodza was of how the crucible of the First World War remade Europe at the same time as it made agrarianism relevant, he reflects on the failures of the system that broke down in 1938:

> After lasting for twenty years, from 1918 to 1938, the Central European order as set up by Versailles is partly destroyed, partly altered in its foundations and

partly living under the threat of being overthrown as well. Does it mean another failure? This appearance of a double failure, the one in building up a first primitive stage in the period 1904-1914, and then that in preserving the Versailles solution does not seem actually to have the meaning of failure. We simply have to face the fact which no statesman and no historian and no sociologist should deny. This fact is that a definite shape of Central Europe, comprising a federation of its free nations, has to share all the pains of a new Europe which could not succeed in being created in a pacific way by diplomatic channels, and which has to emerge out of blood and toil and suffering like all great achievements and political settlements.[48]

The first time around, the agrarians did emerge like a phoenix from the ashes of the war. That feat, unfortunately, was impossible from the ruins of the second.

The other useful frame to contextualize the formation and activities of the Green International is the history of its rival, the IPC (*Krestintern*). The activization of the peasantry and the growing role of the agrarian parties in the wake of the First World War drew the attention of the leaders of the *Comintern*, created in 1919. After all, there was one agrarian party in power (Bulgaria) and in three other postwar states, peasant parties participated in coalition governments (Czechoslovakia, Poland, Latvia, and Estonia). As the communists looked at the agrarian movement as a catalyst to accelerate the coming of the revolution, they raised the slogan of a "worker's and peasants' government," a theory developed by Lenin after the experience of the Russian revolutions of 1905–07 and 1917.[49]

The "worker's and peasants' government" was meant to be a provisional one, formed on the basis of a united front between the Communist Party and the reformist peasant, petty bourgeois, and even bourgeois parties in order to effectuate the peaceful transition toward the dictatorship of the proletariat. It is within this vision that the leaders of the *Comintern* decided to create an international peasant organization, which would base its activities on the platform of a united front with the working class and its Communist Party. All of this became possible after the end of War communism and the adoption of the New Economic Policy (NEP) in 1921.

The decision to create a Peasant International (*Krestintern*) was taken at the Third Plenum of the ECCI (Executive Committee of the Communist International), which held its sessions between June 12 and 23, 1923.[50] The immediate stimulus for this decision came from the events in Bulgaria. The coup of June 9, 1923, had toppled the Stamboliiski government and the Communist Party had taken a wait-and-see stand prompted by the realization of its weakness in the face of a successful military coup. However, officially, its position was formulated as neutrality in the face of an internal bourgeois contest in which the BANU was considered to be a petty-bourgeois formation. The plenum of the *Comintern* criticized the position of the BCP and adopted a resolution compelling all communist parties to strengthen their work in the villages in the name of a "worker's and peasants' government." The *Comintern* created the Peasant International in view of the theory that 1923 saw the beginnings of a second revolutionary wave in Europe.[51]

The formal announcement about the formation of the *Krestintern* came during the International Agrarian Exhibit in Moscow, between October 10 and 16, 1923. While representatives of agrarian cooperatives and agrarian periodicals were invited, the core of the conference was made up of communists, and the (de facto appointed) members of the leading body of the *Krestintern* were all representatives of fifty-two communist parties. The *Krestintern* was modeled closely on the *Comintern*. It was organized around the idea of an international peasant conference, to be held every two years. It published the journal *Krestianskii internatsional* in four languages (Russian, English, French, and German), in addition to the popular series *International Peasant Library*. Each country was supposed to have at least one section, under the control of the Communist Party. Mirroring the structure of the ECCI, larger sections were created for Western Europe, the Balkans, the East, Latin America, and the colonies, as well as a section coordinating the activities of cooperative organizations. Officially the *Krestintern* was to be a nonparty organization, whose main task was to dispel the mistrust of the peasants toward the communists, to penetrate their ranks and establish contacts with their existing political or economic organizations, and to create communist factions in the peasant parties.[52]

In fact, since the *Krestintern* did not have concrete local structures, it was the communist parties that were charged with the implementation of the decisions of the IPC and the ECCI. The propaganda work was delegated to peasant-communists and the question of creating separate *Krestintern* cadres was never put on the agenda. The instructions about the work of the Peasant International were that the communist parties should isolate the Right wings of these organizations, strengthen or create a Left alternative, and introduce the notion of a worker-peasant union, essentially pursuing the tactics of division and schism within the agrarian parties.[53] Understandably, not a single agrarian party adopted the idea of a united front and the leading role of the working class. The Permanent Presence (PP) of the BANU was categorically against contacts with the communists and argued that the united front was detrimental to the agrarian movement, leading solely to the accomplishment of communist goals.

The greatest achievement of the IPC was the creation of a Left wing in the BANU which was able to lead an independent life between 1924 and 1931 in close cooperation with the BCP. The main reason for this was that after the June 9 coup, some members of the BANU saw in the Communist Party the only organization ready to help them in their attempt to overthrow the government of Tsankov. They accordingly sought the cooperation of the BCP through the Vienna Center of the *Comintern*.[54] The participation of BANU members in the September 1923 uprising cemented this cooperation, and it seemed that the Peasant International had achieved in Bulgaria one of its goals: the formation of a united front between communists and Left agrarians. Even after the crushing of the uprising, the communist and agrarian representatives abroad cooperated in aiding the political emigrants, mostly in Yugoslavia and Czechoslovakia.[55]

Even as they were preparing for another armed uprising in the summer and fall of 1924, the cooperation between communists and Left agrarians was fragile, and each group was distrustful of the plans of the other. The misconceived terrorist assault on the "Sveta Nedelia" Cathedral on April 16, 1925, brought about not only the rout of the Communist Party; it provoked a wide-reaching White Terror that decimated the communist leadership and the leadership of the Left and Centrist wings of the BANU.

After 1925, there were two centers in the Representation Abroad of the BANU, as the executive organ of the party outside Bulgaria was called in contrast to the domestic Permanent Presence of the BANU. The first, led by Obbov and Kosta Todorov, was against the united front with the communists; the second, under the leadership of Nedialko Atanasov and Hristo Stoianov, while critical of the dictatorial attitude of the *Krestintern*, did not renounce the links to the communists and continued to receive financial help from the IPC until the end of the 1920s.[56]

The relations were further damaged by the official adoption of the theory of social fascism, which was transferred also onto the agrarians, by the Tenth Plenum of the ECCI in July 1929. All agrarian parties that refused cooperation with the communists were labeled as agraro-fascists. After the faction of BANU "Vrabcha 1" entered the People's Bloc and became one of the governing parties in 1930, BANU's independent Left wing ended its existence. A few of its members joined the Worker's Party, the legal face of the Communist Party, but most merged with the centrist BANU "Vrabcha 1."

The agrarian parties in Yugoslavia were equally skeptical toward the Moscow-based Peasant International. Most characteristic is the attitude of Stjepan Radić. His rapprochement with the *Krestintern* and the ensuing reaction in Yugoslavia is followed in detail in Chapter 5, but here I would like to see it from the point of view of the *Comintern*. Representatives of the *Comintern* got in touch with Radić after his failed attempt to secure help for Croatian autonomy from London. On his way back, in Vienna, Radić was visited by E. S. Goldstein, a secretary at the Soviet Embassy and an agent of the Viennese Bureau of the *Comintern*, who offered Moscow's help in Radić's struggle.[57]

In June 1924 Radić arrived in Moscow, and on July 1, 1924, the CRPP became a member of the Peasant International, citing the closeness of the aims of both organizations on all major issues, such as the land question, the rights of workers to nationalize factories, the liberation of national minorities, the organization of the Yugoslav federation, the formation of a Balkan federation.[58] Nevertheless, when on his way back from Moscow, Radić was met by Maček in Vienna, the latter asked him what he had achieved. "Nothing," Radić's responded, "the communists don't expect allies, only servants."[59] According to Biondich, "His trip to Moscow was designed simply as a means of attaining a measure of international legitimacy, which might in turn strengthen his bargaining position vis-à-vis Belgrade."[60]

All subsequent efforts of the IPC to press Radić to send his representative to Moscow and to initiate a Balkan peasant conference remained unanswered. By December 1924, the leadership of both the *Comintern* and the *Krestintern* realized that the relations with the CRPP were hopeless and that for Radić the link with Moscow was a liability.[61] By mid-1925, when it became clear that Radić was a lost cause, the IPC started a campaign to split the CRPP with the help of the Communist Party of Yugoslavia. The results were negligible.[62]

The small Serbian agrarian party (*Zemljoradnička Stranka*) of Radosav Đokić, on the other hand, hoped that the links with Moscow would strengthen its positions both internationally and financially. Arriving in Moscow in May 1925, Đokić participated in several meetings with the IPC but after his departure, the relationship with the *Krestintern* waned, mostly over the refusal to give him a solid financial subvention.[63]

The interaction of the *Krestintern* with the MAB, and thus by extension with the Republican Party of Czechoslovakia was one of failure and hostility, despite some initial efforts of the MAB to reach out to Moscow. The unwillingness of the *Krestintern* to participate outside of a leading role meant that from the late 1920s, the ideological animosity between the two organizations intensified, with the MAB being dismissed as a kulak formation. It was in this context that the anti-Bolshevik imaginary found its real elaboration.

In the meantime, the *Krestintern* entered a period of decline, fostered by the competition with the rising Green International. On January 12, 1931, a special commission decided to wind up the activity the Peasant International on the pretext that it had not managed to become the international center of the peasant movements and that these movements in the capitalist countries and their colonies were weak and with little prospects for growth. That same commission shifted the task of mobilizing the peasantry to the communist European Peasant Committee in Berlin, but that bureau conducted its last congress in 1932. While the *Krestintern* was not formally dissolved until 1939, the priorities and realities of Stalin's collectivization scuttled it.

The ideas from Bulgaria and Czechoslovakia about the function of the Green International differed even initially. After the possibility for decisive Bulgarian input was truncated with the overthrow of the BANU government, the Green International developed to a large extent according to the directives and interests of the Czechoslovak Republican Party. Yet even as this institution continued to be recast throughout the 1920s and 1930s, as at its inception, this occurred in relation to the two decisive frames I treated above: the international situation and the competition with communism. For the sake of completeness, the Green International also differentiated itself from what it saw as the nonpolitical, professional organizations of the capitalist system such as the International Institute of Agriculture in Rome or the proposal for an international confederation of agricultural organizations championed by Ernst Lauer, but the challenge coming from this direction was minute in comparison to the threat from the *Krestintern*.[64]

This chapter has mined the competing visions of agrarian cooperation that complicated the formation of the Green International in the 1920s in order to situate this agrarian internationalist initiative within the geopolitical context following the First World War. The influence of rival organizations and ideas demonstrates the multiple points of contact which linked agrarian initiatives to major trends in European history. Rather than the peripheral and ineffectual institution as it has heretofore been portrayed, the Green International emerges as a focal point that can serve as a foundation for the reexamination of the international order after the First World War. The next chapter develops the idea that a recasting of agrarian history can better elucidate the agrarian past and at the same time contribute to and inform broader theoretical and empirical subjects, including the study of nationalism. Through a close reading of the orthographic debate in the early 1920s, the next chapter details the competition over the definition of citizenship, nuances the dominant framework for the description of nationalism in East European studies, and interrogates the relevance of subaltern studies to the region.

# 3

# Reimagining the Nation

## Nationalism in the context of agrarianism

*The National Question in Yugoslavia: Origins, History, Politics* by Ivo Banac enjoys the status of the definitive monograph on the subject. Especially for political science or policy purposes, Banac's serious scholarship continues to provide a stable foundation. The position that this monograph occupies is deserved, yet its framework is no longer satisfactory. This statement does not mean that its scholarship is faulty; in Yugoslavia, and in particular in Croatia, the language of nationalism does predominate. However, Banac's genetic approach overstates the case and obscures other factors and inputs. In the preface to *The National Question in Yugoslavia*, he succinctly characterizes his project:

> This is, in short, a genetic study, which traces and analyses the history and characteristics of the South Slavic national ideologies, connects these trends with Yugoslavia's flawed unification in 1918, and ends with the adoption of the centralist constitution of 1921. It will not take readers through the labyrinthine byways of the two interwar decades, nor will it guide them through the dramatic wartime struggles and the postwar socialist experiment. But none of these developments, each in its own way connected with the new twists in the national question, can be understood without the prelude that contained all the seeds of future disorders.[1]

This ingenuous characterization is a testament of the times in which it was made and fits squarely within the schematic and typological tradition in which nationalism was an artifact to be isolated, dissected, and classified.[2] Three decades later, however, that statement sounds predeterministic, an essentialization of the nature and mechanisms of nationalism which is related to the static reductionism imbedded within the civic/ethnic antinomy. The purpose of the second part of this chapter is to provide an alternative reading of the relation and function of nationalism to the agrarian movements in the interwar period.

I would like to begin with a disclaimer: nationalism is not derivative of the ideology and politics of the agrarian parties. Just as these parties were a product of modernity and strove to articulate and institute a more equitable modernity for the peasant masses, so was the concept of the national community integral to the actualization of that project, at the bare minimum through the institutions of the nation-state (though not necessarily the unitary one). That said, thus providing a common denominator, the expression of nationalism varied depending on the tactical choices and strategic considerations that had to be taken to advance the agrarian project. What I mean by this is that in a conjuncture in which an agrarian project could be compromised by the expression of a revanchist nationalism, as was the case in Bulgaria, claims in the name of the nation against an external "other" were muted. Conversely, in a centralizing structure such as that of interwar Yugoslavia, discursively, the national question was amplified because of its political utility in creating the space which was perceived to be necessary to institute the agrarian program in the well-known Croatian case. However, the contingency of this expression is underscored by the diametrically opposed relation to nationalism that was expressed by the *Savez Zemljoradnika*, which perceived its inclusive program of peasant emancipation to be threatened by fractioning along national lines.

I want to make clear that if we overemphasize nationalism as a political movement, as subaltern studies warn us against, much of the complexity of the sociopolitical function of nationalism is occluded. This does not mean that a parallel category such as the "subjective realm" in the parlance of Chatterjee is necessary in the case of interwar agrarian Europe. All that is needed is to pay attention to the contingency of national expression that occurs on the surface of any ideology and praxis. The fact that the variability of national expression can occur not as a result of variability in, for example, agrarian ideology and praxis but even over a stable structure as in the Bulgarian case of the 1920s is an even better demonstration that expressions of nationalism can and sometimes need to be decoupled from iron logics of causality from deep structures. It is precisely this conditionality that frees the analysis of nationalism in Eastern Europe from essentialist genetic reductions and historicist determinism.

After the Second Balkan War and during the First World War, the virulence of Bulgarian nationalist rhetoric against its Balkan neighbors was at its height. That a politician of Stamboliiski's stature could oppose this rhetoric in these circumstances of nationalist hysteria was noteworthy. That he could go even further and invert the discourse is symptomatic of the alternative articulation of nationalism that is the topic of this chapter. Two quotations reflect Stamboliiski's complex relationship to nationalism. The first was pronounced on the floor of parliament when discussion turned to Austria-Hungary's invasion of Serbia in August 1914. Stamboliiski's announcement of his hope for a Serb victory was met by jeers and charges that he was a *surboman*[3] and traitor. To this Stamboliiski famously retorted, "At a moment, such as the present one, when our South Slav brothers are threatened, I am neither a Serb nor a Bulgarian, I am a South Slav."[4] To this day, this pronouncement cannot be forgiven by Bulgarian nationalists and regularly resurfaces as proof that Stamboliiski was a national traitor in these circles.

In a similar vein, Stamboliiski resisted the expression of xenophobic nationalism when he argued for the release of fellow political prisoners in 1918 with King Boris. Nikola Genadiev from the National-Liberal Party was incarcerated during the war for his resistance to Bulgaria's entry into the war on the side of the Central powers; he was found to have been involved in the des Closières affair in which the Entente had tried to draw Bulgaria to its side through substantial grain purchases above market value. When his name came up, King Boris indignantly asked, "But do you know what kind of man Genadiev is?" Stamboliiski's answer that Dr. Genadiev was one of the foremost Bulgarian politicians did not satisfy the king who qualified, "He is not a Bulgarian, but he was produced by a Greek." This time too, Stamboliiski's answer sidesteps national othering:

> I do not know this, but even if I did, that would absolutely not prevent an unbiased assessment. There is a mass of Bulgarian politicians with pure Bulgarian blood, of course if we can even speak of purity in this blood, which are absolutely not suited for their position and purpose. They are simply a superfluous and harmful ballast with their sheepishness, cowardice, and servility in the political life. It appears to me, that this terrible vice is absent from Genadiev and therefore he deserves a different consideration.[5]

Too much attention to anecdotal examples, such as the ones above that I included to establish a tone, has been combined with an excessive emphasis on Stamboliiski's rejection of irredentist goals and foreign policy initiatives by his detractors in order to advance the argument of national betrayal. Even moderate and sympathetic treatments such as those of John Bell refer to an "anti-national" foreign policy that is decoupled from the transformative domestic reforms.[6] This limited view of nationalism obscures the intimate relationship between these two projects.

It is insufficient to characterize this relationship as a sacrifice of Bulgaria's interests on the international stage in order to secure the space for the implementation of domestic reforms, although there certainly is an element of this. Rather, the consistent policy to erase the confrontational relations with Bulgaria's neighbors was an effort to recast and reimagine Bulgaria's interests and thus to transform the nation in synchronicity to the internal national transformation. I would like to recall here two initiatives to clarify the recasting of nationalism. The two proverbial institutions that accomplish the socialization process producing the nation are the army and education. Stamboliiski's two initiatives, the labor service (*trudova povinnost*) and the education reform are crucial to an understanding of what type of nationalism was being formulated in the early 1920s.

The Compulsory Labor Service replaced compulsory military service with the requirement that male and female youths work for the public good. Its ambitions had to be diluted because in the eyes of the Allied Control Commission, it looked like an attempt to avoid the disarmament provisions of the Treaty of Neuilly. However, even in its more diluted form, the socializing and educational component remained on par with its economic value. In propagandizing for the service, Stamboliiski emphasized both the increase of productivity and well-being that would result from it and the development in the Bulgarian citizen of a sense of responsibility and care for public property.[7] The formation of the new Bulgarian subject was continued in the educational sphere. The BANU expanded school construction, made secondary education compulsory, and revised the curriculum. If one adds to this the formative impact of the cooperative movement and its intention of economic liberation, then one can see very clearly a multipronged attempt to engineer a new Bulgarian citizen.

Stamboliiski's non-confrontational foreign policy with Bulgaria's neighbors is even more critical as an illustration of the subordination and transformation of nationalism with regards to the domestic program. In other words, the recasting of nationalism operates through a relational principle in which the domestic reworking of the nation produces a non-irredentist expression of Bulgaria's national interests that through a feedback loop reinforces the primacy of the principle of social equity and cooperation within the nation and among nations. The coup against the BANU in 1923 truncated this experiment and restored the irredentist pattern of Bulgarian national expression and the mentality of encirclement in which foreign policy was the vehicle for fixing Bulgarian nationalism against the external "other." To a certain degree, then, the potentiality of Stamboliiski's efforts to recast Bulgarian nationalism remains in the realm of the hypothetical or counterfactual because the agrarian experiment was not accorded enough time to mature and come to fruition. However, the very fact that such an alternative was elaborated is important as a precedent because, as this monograph argues time and time again, the agrarian alternative was so important that all the other political currents in the interwar period had to respond to and position themselves against it even in its rejection. This, in other words, is the manifestation of the principle that the "failure" of the agrarian experiment cannot belong to the dustbin of history but is indispensable for the proper comprehension of the interwar period.

## The case study

The significance of language and literature as formative for national identity is a major trope within studies of nationalism. The act of imagining and realizing a nation and a nation-state along these lines has been tackled from multiple angles, including folklore, linguistics, and historical studies of educational institutions and their curricula, with a prominent place reserved for textbook analysis. Even though historical studies for the most part have underscored the rapidity and novelty of these processes, a temporal claim that is part and parcel of the larger and fundamental assertion that nationalism is a modern phenomenon, these processes are negotiated over a period of decades. In Braudelian terms, the standardization of a national language would fall somewhere between *conjoncture* and *événement*.

Political life is rarely sensitive to long-term developments in the *structure*, yet within the other two tiers it resolves to *power relations*, social and/or individual. It is this idea of *power* and the related concept of *legitimacy* that constitute the other critical pillar of nationalism studies. Constructivist and modernist conceptions of nationalism illuminate the changing power relationships between individuals, collectivities, and the state within the nationally imagined community, as well as the institutional developments that make these relationships possible. In the intermediate plan, literary-linguistic developments and national-political projects are linked in a knowledge-power relationship. By looking at nationalism through the intermediate lens of the conjuncture, its reification into a static structure and the overemphasis on the rapid vacillations characterizing politics can be avoided.

In the modern period, nationalism has spanned the gamut from an emancipatory, revolutionary, oppositional, and even internationalist ideology, to an exclusionary, official, and dominant one. Explanations for this variation, however, have too frequently been static. The tendency to equate nationalism only with its extreme manifestations not only produced the problematic binary of nationalism versus patriotism but also seriously retarded the development of nationalism studies by treating the two as separate phenomena, only one deserving explanation and censure. The subsequent classification of nationalism into a civic and ethnic type resuscitated identity politics involved in privileging the Northwestern European and American cases. In its most extreme, it cemented this difference by imputing the operation of deterministic logics to each type. In both cases, the dynamics between language and politics on an intermediate scale are lost and the problem is compounded by the accretion of normative and essentialist notions about one type of nationalism against its "other."

In its treatment of the conflict over the BANU orthographic reform, this chapter begins by tracing the correspondence between the conjunctural chronology of changes in Bulgarian orthography and the mutable nature of Bulgarian nationalism while opposing this inquiry to the continuing influence of the civic/ethnic antinomy on the scholarly literature of Eastern Europe. Showing that orthographic reform was an important element of the discourse of ethnic as well as civic nationalism not only fractures notions of structural continuity predetermined by the path to a nation-state, but is exemplary of the fluidity of national identity as it interacts with parallel discourses of modernity.

The orthographic reform reimagined the national community and in a significant, structural way paralleled a shift in the organization of the nation. Simply put, it was intended to and eventually succeeded in recasting the role of the Bulgarian citizen. The content of political integration and the emphasis on a literate citizenry backed by a democratic spelling system could be represented as a drastic shift from an ethnic to a civic conception of the nation. Following Yuval-Davis' emphasis on multiple and contradictory conceptions of the nation competing within a given polity, the caveat follows that the expansive cultural enfranchisement of the Bulgarian speakers in the interwar period did not involve similar initiatives toward the minority populations, especially the Muslims.[8] More importantly, however, the debate undermines this very duality within the workings of Bulgarian nationalism. More appropriate is a vision of nationalism that accommodates progressive and traditionalist inputs; the balance of the two tendencies shifts according to various sociopolitical pressures. Most importantly, it is this dialectic, rather than a predetermined path, that predominates and provides the theoretical basis for more rigorous analyses of nationalism in Eastern Europe.

The agrarian movements of Bulgaria and Croatia in the interwar period receive a very different treatment in the literature with regard to nationalism.

Whereas the Bulgarian case has been described as anti-national, the Croatian one, on the other hand, is represented as the embodiment of a national movement. The utility of this polarity, however, is rather circumscribed as it ossifies and essentializes the complex relationship of the agrarian movements to nationalism and the imagined community at a moment when the character of that community was being transformed through the introduction of the peasant subject. Thus, the second part of this chapter will illustrate how the different outward expressions of nationalism in the Bulgarian and Croatian agrarian movements are in fact based on very similar conceptions of the peasant right and the peasant state. Theoretically, while the first half of the chapter is a critique and response to the distortions of the civic/ethnic binary, the second part elaborates an opening in the treatment of Eastern European nationalism that draws inspiration from the work of the Subaltern Studies Group. The critical juxtaposition of Eastern European agrarian nationalism to the emancipatory project of subaltern studies accentuates its radical merger of the nation, the peasant state, and the peasant subject. Again, as in the first part of the chapter, a dialectic is present that produces polar "national" expressions based on the perceived hurdles to the establishment and defense of the peasant state.

## The theoretical problematic

The beginnings of a systematic exploration of nationalism that transcended the available ethical and philosophical inquiries or the triumphalist histories of national historians can be traced back to the period between the two world wars in the early work of scholars such as Louis Snyder, Carleton Hayes, and Hans Kohn. The methodological imperative to classify or categorize nationalism before a general definition or theory was generated, coupled with the need to account for its destructive potential, generated several taxonomies in the years immediately following the Second World War. While some works, such as Snyder's typology in *The Meaning of Nationalism* adopted a diachronic categorization that relegated the period of "aggressive nationalism" to 1900–45, others advanced frameworks that placed extreme manifestations of nationalism alongside more innocuous ones.[9] In the work of Hayes, "integral nationalism" is the extreme, centralized (fascist) kind listed alongside the "humanitarian," "liberal," and "economic" types.[10] The most enduring of the early typologies, however, divided nationalism into two kinds: a Western and an Eastern. In *Prophets and Peoples*, Kohn normalized and contrasted a positive Western type arising out of the Enlightenment and stressing individual liberty to a reactive, authoritarian Eastern type.[11] According to Kohn, the differences between these nationalisms were serious enough to produce two divergent forms of the nation. The Western nation was a voluntary union of citizens whereas the Eastern nation was centered on an irrational and exclusivist conception of the *Volk*.

The persistent attractiveness of Kohn's model cannot be explained only in recourse to its political utility in describing first the German *Sonderweg* and then in buttressing the divisions of the Cold War. Its Manichean division taps into notions of "good" and "bad" nationalism arranged along the axis of inclusion/exclusion. Of course, the experience of decolonialization shifted the theoretical focus of the literature on nationalism. Not only did the burgeoning nationalist movements demand more sophisticated theorization, but the linkage of nationalism to modernity first via modernization theory and then through its critique prioritized functional analyses over normative typologies. Through the focus on nation building, scholars such as Tom Nairn, Ernest Gellner, and Benedict Anderson explored economic, political, and cultural transformations within a grand theoretical framework that implicitly avoided a positionality vis-à-vis Kohn's binary.[12] The explicit theoretical dismantling of Kohn began from the late 1980s when the theoretical literature on nationalism itself underwent a reevaluation. The deconstructions that accompanied the epistemological crisis in the social sciences and coalesced in the postmodern critique problematized the grand narratives from a multiplicity of angles such as gender, the subaltern, or the production and reproduction of national identity. Currently the most interesting work examines nationalism as a discursive field or practice. One would be hard pressed to find a place for Kohn's antinomy in Étiene Balibar's treatment of nationalism as a "construction whose unity remains problematic, a configuration of antagonistic social classes that is not entirely autonomous, only becoming *relatively* specific in its opposition to others."[13]

One would, thus, suppose that Rogers Brubaker's famous pronouncement in relation to the primordialist/modernist debate that "no serious scholar today holds the view that . . . nations or ethnic groups are primordial, unchanging entities" could be extended to cover essentialist divisions of nationalism into two mutually exclusive types even within the modernist category.[14] The few exceptions such as Leah Greenfeld's *Nationalism*, which will be analyzed in greater detail below, do indeed prove the rule.[15] However, if the civic/ethnic antinomy has by and large been superseded in the theoretical literature on nationalism, it is still firmly entrenched within Eastern European studies.[16] Even if we leave aside the excesses of the largely journalistic employment of the "ancient hatreds" thesis, it is fair to say that the dominant paradigm is one of exclusive versus inclusive nationalism. There are two collective synthetic volumes on nationalism in Eastern Europe, both edited by Peter Sugar.[17] The first (1969) distinguished between aristocratic, bourgeois, popular, and bureaucratic nationalisms. In the second (1995), Sugar had "updated" his framework by employing Kohn's division.

Criticism of nationalism in Eastern European studies is in its infancy. It is telling that Jeremy King declared war on "ethnicism," the bleeding through of primordialism in Eastern European historiography, in 2001.[18] Innovative approaches to nation and state building have begun to make important strides. Irina Livezeanu's *Cultural Politics in Greater Romania* and Timothy Snyder's *The Reconstruction of Nations* are worth examining for their successes and shortcomings.[19] Livezeanu's treatment of cultural politics in the unification of Romania is extremely useful and follows the shift toward cultural analyses of identity formation. On the other hand, her work attempts to explain the radicalization of Romanian nationalism before the Second World War and its exclusivist ideology, a problematic that implicitly perpetuates the inclusive/exclusive binary. Snyder also adopts an innovative position with respect to the standard periodization of Polish history. While his historicization of nationalism parallels the ethno-symbolism of Anthony Smith, he too adopts the inclusive/exclusive dichotomy and focuses on the gradual elimination of the former. Attention to the variable construction and interweaving of identities remains strongest but somewhat isolated within the anthropological literature of Eastern Europe.[20]

What explains the disconnect between the advances in theoretical works on nationalism and the persistence of the civic/ethnic antinomy in Eastern European historiography? For one, at the heart of the translation of civic and ethnic categories from contextual specificities to ontological truths is the embedding of the antinomy within the larger power discourses that divide Eastern Europe from the rest of the continent. The layering of the binaries modern/backward, individual/collective, democratic/autocratic, developed/undeveloped continue to allow the civic/ethnic antinomy to creep back into analyses so long as essentialist approaches to the treatment of variation (East/West) remain firmly entrenched within Eastern European studies. The other contributing factor derives from some of the thematic tropes of Eastern European historiography. The traditional attention to wars, irredentas, or minorities predisposes toward this distinction. More innovative topics such as the focus on alterity, borders, collaboration, and retribution are welcome developments and the best examples consciously try to bridge the East/West divide.[21] However, they too tend to focus disproportionately on moments/cases of extreme tension. The recent trend in rehabilitations of the multiethnic empires is itself predicated on a contrast with the exclusivist Eastern European nation-state model.[22]

Clearly, the effectiveness of theoretical objections to the civic/ethnic antinomy is limited. It is so entrenched within Eastern European historiography that it is worth scrutinizing it on its own terms. In order to illustrate the problems with the dichotomy, it is worth focusing on Greenfeld's *Nationalism*, its most sophisticated, recent version. The success of the book, based to no small degree on its scope and the author's erudition, is also firmly rooted in the neat modeling that it proposes. To be fair, Greenfeld is very careful to qualify and moderate her claims.[23] However, two disturbing propositions lie at the heart of her inquiry and radically contradict her measured assurances. The first can be described as an engine of alteration that corrupts nationalism at the time of its spread, often under the pressure of Nietzschean *ressentiment*. The result is an inverted or negative telos redeemed only by the emergence of the American national polity.

The second proposition encapsulates the essential flaw of Greenfeld's argument because it establishes the internal logic that generates this hierarchy. It is the equivalent of a statement of Original Sin, in privileging the initial formation of national identity and disregarding its subsequent spread: "The character of every national identity was defined during its early phase . . . Its effects, in the political, social, and cultural constitution of the respective nations, as well as their historical record, are attributable to this original definition which set the goals for mobilization, not to the nationalization of the masses."[24] The deficiency of such approach is obvious: it leads to essentialism and ahistorical modeling. The conceptual rigidity also robs nationalism of its most significant characteristic, namely the ability to reinvent itself and remobilize the population as a response to *specific* historical situations. In the end, nationalism resembles a desiccated archaeological artifact instead of the living social construct that it really is.

On the surface, Bulgarian nationalism seems to fit easily within ethnicist categorizations. The nationalization of Bulgaria was elite driven, heavily grounded in cultural initiatives of which language was one of the central pillars of national identity and was intimately entwined with ideas of historical continuity. It is a case of nationalism imagined along the lines of ethno-linguistic continuity. It was also formed against the Greek and Turkish other.[25] If we follow Greenfeld's logic, this means that *ressentiment* is the deep structure of Bulgarian nationalism.

In order to transcend such a static portrait, it would be useful to insert a distinction between *notion* and *concept*. The crucial difference is that *notion* does not clearly mark the contours of the object it describes and thus its utility lies in its function as rough synthetic and intuitive knowledge. While Greenfeld identifies obvious variations in the way nationalisms have developed, in my opinion she is culpable of extending her notional insights into conceptual categories. If all developmental paths to the modern nation-state can be conceived of as being united within the larger context of modernity, then forcing a hard categorical distinction into two types based on *ressentiment* produces precisely the conceptual essentialism that is at the heart of the civic/ethnic antinomy.

## The centrality of the language question

When the Bulgarian revivalists confronted the issue of the formation and standardization of the literary language in the nineteenth century, their initiatives were marked by the burden of history. The perception, content, and meaning of the historical burden changed over the course of the Bulgarian Revival[26] and it exerted a profound influence on the choices of grammatical or orthographic structure of the language. That crucial link had been forged in 1762, when the monk Paisii Hilendarski advanced the idea of language as the expression of national consciousness. The famous exhortation of his *Istoriia slavenobolgarskaiia* for the Bulgarian to remember his glorious past and not be ashamed of his language set the tone for the interrelatedness of nation and language.[27]

Given the absence of a powerful center, which could impose a particular dialect, it was natural for the language of the Middle Ages to exert tremendous influence. Indeed, in contrast to his radical national ideas, the linguistic choice of Father Paisii fell on the conservative side as he based his work on Church Slavonic. The other literary tradition—that of the Damascenes, emerging from the sixteenth century—covered texts in the vernacular of religious edifying content. Whereas Church Slavonic continued to use a case structure, the vernacular had switched to a synthetic (analytical) grammar already before the Ottoman conquest.[28] Not only were the Damascenes more accessible to the population, but they offered the possibility of solving the language question with greater radical potential for the advancement of national and literary aims in tandem.

In the 1830s and 1840s, these traditions crystallized into two competing approaches to the problem of codifying and standardizing the language. The

New Bulgarian school insisted on the vernacular, while the Church Slavonic School elevated the language of the Orthodox Church, the institution that, as the argument went, had preserved the "nation" through the centuries of the "Turkish yoke." A third school that emerged a few years after the others, the Slavo-Bulgarian, occupied the space between the two and sought a compromise position by suggesting an analytical language based on the common elements of the Bulgarian dialects with Church Slavonic. This position was similar to the work of Vuk Karadžić and Adamantios Koraïs in the Serbian and Greek contexts, respectively.

Scholars frequently refer to the triumph of the New Bulgarian school by the middle of the nineteenth century, when its prescriptions are seen as having set the direction for the future development in morphology, phonetics, and the lexicon.[29] This interpretation largely holds from the perspective of hindsight, but it also obscures the rather murky reality of that "victory." For one, the New Bulgarian school was less able than its competitors to produce a concrete program manifested in a cogent translation of the vernacular into its literary counterpart. Second, the argument that it accommodated inputs from the Slavo-Bulgarian and Church Slavonic literary traditions in order to flesh out and enrich the language in the latter part of the nineteenth century begs the question. It invites the query of whether, at least structurally, it would be more appropriate to talk of negotiated compromises more akin to the program of Neofit Rilski, the chief proponent of the Slavo-Bulgarian platform.

An egress from this morass is offered by Stoian Zherev when he characterizes the position of the New Bulgarian school not as a concrete victory but as one of principle: "It is more precise to accept that the *view* [original italics] of a literary language with a vernacular basis triumphs. It establishes itself more categorically and more concretely during the movement of the Bulgarian literary language (after the '*Riben bukvar*') from *diatopia* to *syntopia*."[30] Zherev's argument hinges on the simple reality that the linguistic controversies were decided on the stage of the budding Bulgarian press. There were numerous proposals for the Bulgarian language that were only loosely united under the flag of a vernacular basis for the modern literary language. In this context, Zherev provides another clue with his reference to the *Riben bukvar*. Petur Beron's work was the first Bulgarian primer, used extensively in elementary schools. Beron was also one of the chief proponents of the New Bulgarian school. He and Vasil Aprilov, the founder of the first Bulgarian secondary school, were able to utilize their fame and wealth to provide an invaluable nimbus of legitimacy to the New Bulgarian argument. In other words, the triumph in principle of the New Bulgarian school was pushed through due to the exceptional cultural and social capital that its proponents were able to muster around its cause.

This is an example of the interconnectedness of the national and literary projects, where the very form that the Bulgarian language was to take was negotiated in reference to the historical legacy of the nation. Clearly the nexus between the glories, past and future, of a Bulgarian state and its language figured in the minds of the revivalists. But they were not only men of letters. Major revolutionary figures combined the literary and the political into a joint program for Bulgaria's independence. Georgi Sava Rakovski was one of the fathers of the revolutionary movement, the founder of the Bulgarian Legion in Belgrade. He had participated in a revolt in 1841, passed through France, fought in the Crimean War, and lobbied for support for the Bulgarian revolutionary movement in Belgrade and Athens. Rakovski maintained that language was the most important marker of ethnicity and the nation, and that the native language was the chief weapon in the struggle for national progress. His views are reflected in his poetry, but he advanced his literary agenda more directly as editor of *Dunavski lebed*, in which he agitated against the normalization of a single dialect as the basis of a literary language.[31]

Hristo Botev, a prominent member of the next generation of revolutionaries, also trod the dual cultural and political path toward the nation-state. He is as much remembered for his organizational activities in Bucharest in the 1870s as for his talented and frenetic editorial activity and versification. Before he fell in the April Uprising of 1876 at age twenty-seven, he had edited three newspapers: *Duma na bulgarskite emigranti* (1871), *Zname* (1874–75), and *Nova Bulgariia* (1876). *Zname* had a section "Literary News," where Botev reviewed the Bulgarian press, and analyzed and critiqued its language.[32]

The preceding survey of the genesis of the modern Bulgarian literary language makes intelligible the debate over the orthography during the second half of the nineteenth century. However, it also allows an important refinement, or at least a qualification, of the Hrochian periodization of the national revival in Bulgaria.[33] In the case of the codification of the grammatical structure of the literary language, it is no longer appropriate to apply a model of unproblematic transmission of the work of the national enlighteners from Stage A to the succeeding stages along a vector whose only reaction to change involves the size of the groups involved. As the complex reality of nineteenth century literary practice shows, the activity of the academic grammarians was quickly sidelined after its initial explosive influence by the larger group of Stage B activists and the preferences of a reading public in Stage C. The choices of publishers, editors, writers, and especially the various groupings of these agents in the periodical press determined the selection of a radical New Bulgarian rather than a conservative and etymological basis for the literary language, a reality that was confirmed by the codification of this choice when it was implemented as the language of state following independence in 1878. Gyllin writes, "Not only during the Renaissance, but also for a long time afterwards grammars lagged behind existing written realities . . . . Thus, in practice, the modern Bulgarian literary language as it existed in the decades after the liberation had been formed not by grammarians but by those actually using it."[34]

The importance of cracking a teleological account of the national revival is that it permits the entry of politics into the analysis. Rather than a passive public, we are confronted with agents that take over the process at certain moments. The debates over the democratic or traditional basis of the language did not occur in an ivory tower; they were contested in the body politic. One of the sites of public involvement was the *chitalishte*, the reading room for the general public that was widespread in the Bulgarian towns and villages. Especially after the formation of the Bulgarian kingdom and during the early course of its development, there was a growing tendency toward greater social involvement.

By the middle of the nineteenth century, interest in the Bulgarian past had produced a considerable body of literature and the appropriate specialists. The expansion in knowledge changed the balance of forces between traditionalists and modernizers in favor of the former because it became increasingly difficult to ignore the "cultural heritage" of the medieval Bulgarian kingdom. This shift is well represented by Aprilov's publication "Thoughts on the Contemporary Bulgarian Studies" (1847).[35] According to Gyllin, this paper established "certain fundamental historical linguistic facts, which strongly favored traditionalistic views: the difference between Old Bulgarian and Russian Church Slavonic, the direct link between Old Bulgarian and New Bulgarian, and not least that Old Bulgarian was the oldest Slavonic literary language, that it had spread outside the Bulgarian borders and had been of importance for the formation of other Slavonic literary languages."[36] Thus, even though Aprilov continued to reject an Old Bulgarian basis for the language (the historical continuity between Old and New Bulgarian was yet another argument for accepting the vernacular position), his research helped bolster the traditional, or etymological, schools.

While grammatical disputes were raging between the 1820s and the 1840s, the orthographic question had remained on the back burner. Writers not only were free to choose their spelling, but they frequently employed different alphabets, a variation that had twenty-four to thirty-eight letters at either spectrum. But with the grammatical basis of the language settled, the morphological and phonological streamlining that remained to be done was picked up within a broader debate over the orthography. Characteristically, two schools emerged that defined the battle lines in their relationship to the historical legacy. The Plovdiv school of Naiden Gerov and Ioakim Gruev favored a strongly etymological orthography. The Turnovo school of Ivan Momchilov and Nikola Mihailovski retained traditional elements, but a juxtaposition with its competitor clearly reveals its proximity to the phonetic principle. On the one hand, the Turnovo school preserved the traditional placement of the letters ѣ, ь, while on the other, the ur/ru and ul/lu variation reflected the vernacular pronunciation. By the end of 1870, two other schools had joined the fray. Liuben Karavelov was able to advance an orthography that was even closer to speech through the weighty examples of his novels and journals.

The most influential intervention, however, was made by Marin Drinov, the prominent Bulgarian linguist and historian at the University of Kharkiv. A mixture of the phonetic and etymological principles, although leaning more toward the latter, Drinov's position was published in 1870 in the organ of the Bulgarian Literary Society. The Literary Society, founded in Braila, Romania, in the previous year, sought to "spread general enlightenment among the Bulgarian people and to show it the way to its material progress."[37] Eventually, this national emancipatory institution would be transformed into the Bulgarian Academy of Sciences (1911), but even in the decade prior to 1878, it was the most prestigious Bulgarian cultural center. Drinov's proposal exerted tremendous influence. Drinov considered that "the most correct path to the achievement of a consensus on the orthography ... [would be], first, to fully use the living Bulgarian speech in all the varieties found in all the various regions of Bulgaria, and, second, to not neglect the Old Bulgarian in its oldest and most orthographically [correct] monuments."[38] The precise content of his orthographic proposal is not relevant to this chapter, aside from mentioning that he got rid of the Greek letters Θ, Ω, Ξ, and so on, but kept several traditional letters from Old Bulgarian. One of these, the ѣ, served as a bridge across the dialectal divide separating Bulgaria along a North-South axis. Its double value could be read as either [e] or [ia]. The result was that Drinov's conciliatory orthography enjoyed an informal dominance from the early 1870s and especially in the first decades of Bulgarian "independence" (from 1878 to the 1890s).[39]

The 1890s were marked by an increased interest in the orthographic question. The normalization of political life under the administration of Stefan Stambolov provided a large impetus toward cultural projects. A commission of experts convened in 1892 with the aim of producing a standardized spelling. The members of this commission that were to play a significant role in the astoundingly vituperative quarrels over orthography in the 1920s—Liubomir Miletich, Alexander Teodorov-Balan, and Benio Tsonev—first cut their political teeth here, so much so that the debates of the 1890s can be seen as a dress rehearsal for the 1921–23 conflict. Although grounded in philology, their proposal paralleled Stambolov's anti-Russian politics and brought the Bulgarian spelling closer to the Serbian. Strong public reaction against the proposal doomed its implementation, especially after the fall of the Stambolov regime in 1895.[40] Two more official proposals failed to establish themselves in the following years. All the while, Drinov's almost-two-decades-old system continued to attract the most adherents. It is not surprising then that Bulgaria's first official orthography of 1899 affirmed what was de facto already the case. Todor Ivanchev, the Minister of Education at the time, based the official orthography on the proposals of Drinov, Momchilov, and Karavelov, producing a moderate program. Even though it contained a firm phonetic basis, it retained some traditional elements that caused it to be difficult to learn and master. Known as the Drinovsko-Ivanchevski orthography, it remained firmly entrenched until after the First World War.

A seeming paradox emerges from this brief narrative of the standardization of modern Bulgarian. The grammatical structure of the literary language that triumphed was radical and New Bulgarian, whereas the orthography was based on a conservative, Old Bulgarian etymological principle. As we have seen, the reasons for this divergence are historical and related to changes in the vision of how the Bulgarian nation should relate to its history. The language question was an integral component of national sentiment. It was debated, changed, and codified in a process that paralleled the political and social development of the country. As the essence of the nation, language was also its marker. Conflict over the orthography was political and related to questions about nationalism as well as popular participation and popular sovereignty, for language had to foster inclusion into the state and the national community, and at the same time had to serve as the protector of the Bulgarian heritage. At the heart of the language debates were the related antinomies of the vernacular/etymological and the modern/traditional. These antinomies lent themselves easily to reconfiguration as democracy/oligarchy in the crucible of the crisis in authority following the First World War when the radical ideas of self-determination and popular sovereignty gained ground. Omarchevski's spelling reform of June 1921 that further simplified the alphabet was the product of revolutionary changes in all these spheres and, as such, offers a privileged glimpse into their dynamics.

## The orthography debate of the 1920s

On December 20, 1921, Stoian Omarchevski, the Minister of Education in the cabinet of the agrarian government of Alexander Stamboliiski (1919–23), decided he had had enough of the obstructionism of the University of Sofia over the implementation of his orthographic reform. Twelve days earlier, the cracks in the relationship between the minister and the academic council of the university were evident, as Omarchevski refused to attend the annual holiday of the university that was held in the national theater. To give a sense of the significance of this holiday, the guests included the king, the mayor of Sofia, distinguished academicians, and the cream of the political class. On the other hand, the agrarian regime regarded these elements with warranted distrust as corrupt oppressors of the peasant classes. December 20, then, marked the official commencement of hostilities when Omarchevski issued a fine to the rector of the university, Liubomir Miletich, amounting to a twelfth of his salary and ordered that the university coffers be sealed. The immediate cause for the intervention was that Miletich had made use of the old orthography to publish the official university obituary for his colleague, Professor Medvedev. There were other, more serious issues that could have precipitated an impasse like student tuitions, the holding of positions of public office by faculty members, and even the thorny question of university autonomy. In what chroniclers of the university aptly call the second university crisis, these and other issues also found their place.[41] But it was an argument over spelling that began and lay at the heart of the conflict. More precisely, the university's refusal to fully implement the new orthography passed by Omarchevski in the summer of 1921 and its unwillingness to atone for its errant publications crystallized the already developing animosity between the two parties into an outright refusal to work with each other.

In the first phase of the conflict, the frenzied correspondence between the Ministry of Education and the university established the battle lines and the arguments that would be used. The response of the academic council of December 22 firmly rebuffed the intervention of the minister by arguing that the sealing of the coffers was an unlawful act and that only the council had the right to discipline members of the faculty. On his part, Omarchevski countered on December 23 with an explanation that it was not only the Medvedev obituary that offended but also several other publications, including the yearly report of the university for the academic year 1920–21. Furthermore, Omarchevski sought to convene a meeting of the whole faculty that was to determine if the rector had acted alone or as a representative of the university's interest in flaunting the orthographic directive of the Ministry of Education. The indignant reply of the council (December 26, 1921) contained a refusal to convene the university faculty and punish the rector.[42] It added the barb that Omarchevski should be more concerned with his own actions because he himself had made exceptions to the orthography directives.[43]

In January 1922, the university sought a higher arbiter and appealed to the Supreme Administrative Council to remove the fine on Miletich, again by using the argument of university autonomy. Omarchevski, too, sought measures to weaken his opposition by applying legislation that the BANU had passed prohibiting politicians from holding lectureships at the university.[44] In fact, a good number of MPs and leaders of the bourgeois opposition parties were holding appointments at the university. The leadership of the National–Progressive Party (Stoian Danev, Stefan Bobchev, and Petur Abrashev), the Democratic Party (Georgi Danailov and Vladimir Mollov), the Radical Party (Iosif Fadenheht, Petko Stoianov, and Venelin Ganev), and of the National Alliance (Alexander Tsankov and Dimitur Mishaikov) comprised between themselves almost exclusively the faculty of law. The importance of these figures and their political connections are all the more visible if one remembers that the economics professor Tsankov headed the coup against the agrarian government in 1923 and subsequently became prime minister in the new regime. Thus, when Omarchevski demanded the resignation of eight figures from the faculty of law on February 18, 1922, he was attempting a coup against nothing less than members of parliament.

Omarchevski additionally put financial pressure on the university via the new budget for the upcoming academic year. The budget proposed to cut the professorial posts of the most vocal opponents, inclusive of the rector and the pro-rector, and placed a moratorium on new appointments. The response of the academic council was a warning that this crass unconstitutional encroachment on the internal workings of the university would make future cooperation with the ministry impossible and would force it to cease instruction.[45] This position was affirmed on several occasions in March, when the conflict spilled into the press and took on a more ominous tone with professional societies announcing their solidarity with the university and organizing protests. The unwillingness of either side to back down was illustrated by the cessation of classes on March 11 and the government's answer in its chief organ, *Zemedelsko Zname*, on March 16 that this was to be a "battle to the end."[46] Thus, the decision taken by the general meeting of the entire professorial body on April 1 to completely shut down the university was the only possible outcome in light of how serious the situation had become. The university remained closed until August 19, 1922.

The reopening was the result of a series of negotiations that began at the end of June. Still, the resolution of the university crisis occurred only after Stamboliiski sent Omarchevski on a diplomatic mission to Brazil at the beginning of August. The concessions by the government included rescinding the rector's fine, restoration of the university credits in the budget, and certain changes to the university rulebook following the recommendations of the administrative council. On its part, the university agreed to immediately open and follow the existing law. Although the wording of this resolution was not specific, presumably to save face for the university, it implied that it would diligently apply the new orthography. For, in the course of the university crisis, on March 25, 1922, parliament had passed the "Law for a Unitary Bulgarian Orthography" that included specific punitive measures for transgressions. The sanctions included the shutting down of any guilty periodical publication for a period of one to three months, the firing of administrative functionaries at fault, and a one- to three-month jail time with a fine of up to 20,000 leva for private persons.[47] Furthermore, the university undertook to hold elections for a new rector, deans, and members of the academic council, which it duly fulfilled on September 11, and agreed to replace the private lecturers (politicians) with regular employees. The agreement held until the coup against the agrarian government in June 1923, and the university remained open. Although a few issues remained to be resolved, chief among them student taxes, the resolution of the orthographic battle removed the major bone of contention between the two parties.

The battle between the Ministry of Education and the University of Sofia was much more than a peripheral conflict. Over its course, it involved all major political players and mobilized significant sectors of society, particularly in the capital. The language question, which had uneasily mediated between the etymological and the phonetic principles, was eminently well suited to take on the symbolic role of a litmus test dividing the radical populist conceptions of the national polity that emerged in the aftermath of the First World War, from the paternalistic elitism of the etatist apparatus that had developed in the first decades of independent statehood. In other words, the violent sentiments that the spelling reform aroused in its supporters and detractors were a major component of the struggle for the nation's soul in the turbulent 1920s. The extension of the democratic principle and the location of unmediated popular sovereignty in the masses became even more legible in the democratization of an orthography that would erase the distinction between literate and "illiterate" citizens and the privileges that hinged upon it. It is my contention that while the forces on the Bulgarian Right found it difficult to overtly contain the democratic revolution in their political rhetoric, they found a surrogate for their program in the defense of traditions that preserved social exclusions. Since the language question had been so intimately tied up with nationalism, the change in the official orthography could not mark anything less than a parallel change in the nature of nationalism. As seen in detail below, even rhetorically, the bottom line of the impassioned responses to the new spelling always concerned Bulgarian nationalism.

A focus on the localized struggle between the ministry and the university has the advantage of revealing the technical process of escalation in which recalcitrance over the implementation of the spelling reform turned into a battle for the university's survival. A broader, sociohistorical perspective, however, can represent the university's unwillingness to censor its rector as an effort to retain at its core the role of autonomous defender of culture and the national patrimony against the peasants' meddling. It is this second perspective that better puts in focus the symbolic significance of the orthographic issue and exposes it as an integral part of the sociopolitical struggle for ascendancy that was the legacy of the First World War.

The immediate cause of the agrarian regime's decision to examine the orthography in 1921 was agitation by primary school teachers. In several conferences, the educators expressed their dissatisfaction with the difficulties they encountered in teaching the Drinovsko-Ivanchevski spelling. The biggest cause of concern was that it was so difficult to learn, with all its etymological exceptions, that it invariably ended taking up most of the instructional time. In an appeal to the ministry, a teachers' delegation demanded a simplified orthography so that literary and stylistic exercises could be reinserted into the curriculum.[48] The teacher's position was buttressed by the intervention of the prominent linguists Tsonev and Teodorov-Balan, who had presented a proposal for an orthographic reform at the conference.

Omarchevski's response was to issue order no. 1941, whereby he charged a commission with the task of "reexamining the orthography of the contemporary Bulgarian language and giving its opinion on whether it can be simplified so that it would become accessible and easily assimilable."[49] The concluding report presented on June 17, 1921, was eminently sensible. Its chief points concerned abandoning the several letters with dual value or denoting sounds that in the course of the linguistic development had become mute. The organizing principle behind these suggestions was that every letter should correspond to a single unique sound.

The work of the commission was closely observed and discussed in Sofia. Even as it began its work, a lively debate took place on the pages of the periodic press. Teodorov-Balan and Tsonev as well as the literary critic and historian Alexander Balabanov agitated in favor of reforms. Others advanced the case that a small clique was imposing its agenda contrary to any real linguistic need. The fact that Tsonev received death threats and Teodorov-Balan was publicly defamed attests to the pervading rancor.[50] In this tense context, the commission displayed sensitivity to the developing nationalistic tinge of the debate and attempted to dismantle one of the key objections of the detractors. Its final report reads, "These reforms in the orthography do not damage the Bulgarian national unity because national unity is expressed specifically through a literary language and orthography, which are properly established and universal."[51]

The report was supposed to be provisional, to generate recommendations to be examined by the relevant competent authorities, chief among them the historico-philological branch of the BAS. By no means was it to serve as the independent basis for reform. Still, immediately after its completion and publication, Omarchevski presented it to the cabinet for scrutiny and implementation. In a little over a week, the "Directive for a Common Orthography of the Bulgarian Literary Language" made the spelling reform official. It implemented the commission's recommendations with one notable exception: the reduced vowel would be denoted by the Church Slavonic Ѫ rather than by the suggested hard sign. In light of the way in which the opposition to the reform developed, historians have focused on the arbitrary and authoritarian side of Stamboliiski's personality when providing an explanation for why he overruled the recommendations of Omarchevski and the commission, and preserved the Ѫ. Alexander Velev shows another side to the debate, when he examines the minutes of the Council of ministers and tracks the arguments of both sides:[52]

> G. Damianov [MP] insisted on the preservation of the letter Ѫ because it was an Old Bulgarian letter and it distinguished our language from others. Apparently his chief concern was the differentiation of the Bulgarian from the Serbian language. This view doubtlessly expressed the intimate considerations of the initiators of the orthographic reform, who, with the conservation of this letter, intended to disarm the opposition parties whose claim it was that the reform was pandering to the interests of Belgrade.[53]

Velev's analysis is a useful corrective showing that even the highest circles of the agrarian government were not insulated from concessions to nationalist positions. There were good grounds for the agrarians' concern because the resistance to the orthographic reform was serious. It came via objections over the content of the reform and its improper implementation. It occurred at rallies and demonstrations, through the formation and activities of coalition blocks, and in the treatment of a hostile press. At the first level of commonality, the opposition was united in its political struggle against the agrarian regime. Beneath it, one can isolate a second level of commonality, which is the language of nationalism. The two are almost completely isomorphic with the exception of the Communist and Social Democratic parties. Both recognized the need for reform but criticized it over the irregular and authoritarian methods of its implementation. These parties of the Left did not play the popular nationalist card; rather, their critique of the bourgeois parties hinged on the exposure of the dangerous demagogic and reactionary pandering to nationalist sentiments.

On July 12, 1921, a rally against the orthographic reform was convened in Sofia by the Union of Bulgarian Scientists, Writers and Artists, the Association of Journalists, and the Association of Writers. Alongside its more technical objections, the adopted resolution is quite illustrative of the nationalistic overtones, particularly with regard to the Bulgarian irredenta, that opponents to the reforms found useful to employ: "The reform, so abruptly imposed, having paid no attention to the literary traditions that are so precious to our co-nationals outside the boundaries of the Kingdom of Bulgaria, is extremely damaging to the common Bulgarian national interests."[54]

The Treaty of Neuilly-sur-Seine compounded the irredentist obsessions in Bulgaria after the Balkan Wars. The BANU infuriated revanchist circles both because of its history of opposition to the First World War and its policy of rapprochement with the Triune Kingdom. The neglect of the army, the replacement of the compulsory military service with a civil counterpart, and the rhetorical abandonment of any territorial claims were now aggravated by concessions on the cultural front. As we see here and in the discussions of the ministerial cabinet, the reception of the orthographic reform was tainted by the preoccupation with Macedonia and how to preserve, in the worst-case scenario, at least the cultural links with the "fellow Bulgarians." In this rhetorical context, the representation of the BANU policy as national treason was only a short step away.

The voice of the opposition in the press became ever more insistent throughout 1921, especially with the university crisis. At first the critique of Omarchevski's cultural politics focused on the spelling. Liubomir Andreichin refers to the significant negative reaction that held the abandoned traditional letters and spellings to be "symbols of Bulgarianness" and rose in defense of the national patrimony.⁵⁵ An example is Stoian Romanski's article "Orthographic Crisis" of 1922. He too maintained that the Ѫ, Ѣ, Ъ, and ь were "distinguished signs of Bulgarian writing," and concluded that "[The Drinovsko-Ivanchevski spelling] is a reasonable combination of the ten-centuries-old tradition of Bulgarian writing with the demands of the New Bulgarian language. Because of these qualities it should not be replaced by a new spelling system even if it contains a few obvious advantages."⁵⁶ His is a significant example of the implacable yet measured opposition. Romanski was no extremist, yet, the eminent philologist and ethnographer would play a central role in the restoration of the Drinovsko-Ivanchevski spelling after Tsankov's coup.

Once the university crisis developed, the oppositional newspapers began to print in defense of university autonomy as well—these were the newspapers *Radikal* of the Radical-Democratic Party, *Priaporets* of the Democratic Party, and *Nezavisimost* of the National-Liberal Party. Only *Zemedelsko Zname*, the BANU's chief publication, defended the policy of the Ministry of Education. The newspapers *Zora* and *Priaporets* were even closed down for a while at the end of 1922 because they demonstratively continued to use the old orthography.⁵⁷

On February 26, thousands of people protested Omarchevski's proposed firing of the staff of the Law Department. This action was called by the Union of Bulgarian Lawyers. Students marched as well that day and were dispersed by the police. In March, shortly before the university went on strike, twenty-one societies, institutes, and organizations united to form the Cultural-Professional Committee of Citizens in defense of the university. Aside from professional organizations, it included the BAS, *Naroden Sgovor*, and the Union of Reserve Officers. Dimitrina Petrova correctly analyzes the politics of this formation as a committee for struggle against the government: "The university crisis, and especially its conclusion, reveals the isolation of BANU from academic and some other cultural circles as well as the failures in its efforts to integrate them to the regime."⁵⁸ The bigger failure, however, was the alienation of the nationalist circles. In fact, it was even unable to properly represent and then defend its cultural policy against those that deemed it anti-national.

It is beyond the scope of this chapter to treat all the sites of contestation that developed around the orthographic debate. The proposal of the BAS in 1921, which to a great extent replayed the drama of the university crisis as the "Crisis of the Academy of Sciences,"⁵⁹ and the parliamentary debates over the spelling legislation proved to be significant for two reasons: first, they indicate the breadth of the phenomenon and the commitment it elicited from a vast array of agents; secondly, the discursive positions adopted at these two sites are perfect illustrations of the politicization of linguistics in the context of nationalism. The parliamentary debates revealed the invariably nationalist arguments advanced by all bourgeois parties alongside the more serious objections to the orthographic reform and its encoding into legislation. Even the Communists and the Social Democrats treated nationalism in their critique since it was the only differentiating mark that separated their stance from that of the bourgeois position.[60] It is worth stressing that agrarians, communists, and social democrats were appealing to a different conception of nationalism when they made the claim that an orthographic reform was necessary in the name of the common citizen. It is a nationalism connected to popular sovereignty and the interests of the people. By exposing the narrow, demagogic interests of the bourgeois parties in appealing to a defensive nationalism, they were claiming space for an alternative, populist "Bulgarianness" (*Bulgarshtina*) both conceptually and politically. Of course, the coincidence of interests did not prevent the Communist Party from joining the opposition, just as it stood by on the sidelines in June of 1923 during the coup against the BANU regime.

For all their criticism and denunciation of the methods employed by the agrarians in their cultural policy, the conspirators led by Tsankov availed themselves of the same tactics. Ten days after the bloody coup against the agrarian government of Stamboliiski on June 9, 1923, Miletich, the former rector of the university and chief protagonist of the university crisis as well as the man who bears the chief responsibility for the scuttling of the orthographic project of the BAS, was at the head of a new orthographic commission. Its reactionary conclusion to restore virtually verbatim the Drinovsko-Ivanchevski spelling of 1899 was hardly a surprise. The only difference was the concession to use the ѣ only in cases where the dialectal variance had been preserved, rather than in the strict etymological placement demanded by the 1899 rules. One cannot refrain from reveling in the irony that Tsankov's guide to the orthography of August 1923, entitled *Uputvane za pravopis na bulgarskiia ezik* (Guide to the Orthography of the Bulgarian Language), both syntactically mimicked Omarchevski's effort of 1921, *Uputvane za obsht pravopis* (Instructions for a Common Orthography), and was promulgated in exactly the same way by ministerial decree. The 1920s witnessed a farcical iteration of the tremors centered on Omarchevski. The Left-leaning press continued to employ the 1921 BAS orthography. For example, until 1930 Bulgarian language materials published in the USSR utilized the original recommendation of the Orthographic Commission rather than the slightly altered version backed by Stamboliiski. Linguists continued to agitate in favor of reforms, be they of the Omarchevski or the BAS variety.[61] The climax in 1928 involved nothing other than the employment of the legislature, in the same way that Omarchevski had done, to make the old spelling mandatory. The only difference was the conciliatory gesture allowing authors of a limited number of scientific and literary works to choose their spelling.

While the old Drinovsko-Ivanchevski orthography continued to be used in the interwar period, the radicalism of 1944–45 brought a change. A large commission of writers and linguists began work shortly after the communist dominated Fatherland Front took power on September 9, 1944. The February 1945 decree that derived from its work was an affirmation of the phonetic principle. The orthography legislated in 1945 was basically identical to the agrarian one. While it is more precise to say that its immediate model was the proposal of the historico-philological branch of the BAS, the agrarian legacy is clearly visible in the sociocultural grounds for its implementation.[62]

\* \* \*

What explains the observer's initial awe at the tumult that was caused by a seemingly technical issue, the orthographic change in the 1920s? The answer this chapter proposes is that the spelling reform was much more than a matter of linguistics. It marked a fundamental change in the Bulgarian polity and was associated with a moment of restructuring, democratization, and reshuffling of elites—the rearrangement of norms that favored different social and cultural constellations. It is only appropriate that something so momentous would have been debated in the dominant discourse of the time: nationalism. This discourse incorporated all the anxieties or peculiarities of the historically contingent face of Bulgarian nationalism. There is a direct link between the debate over orthography and the debate over the face of Bulgarian nationalism. But there is also something more involved here. This was not simply the couching of an external debate in the trappings of nationalism. The centrality of the linguistic base for the nation (the ethno-linguistic community) in Eastern Europe made this a contest for the soul of the nation.

The cultural/political importance of orthographic reform was its attempt to open the polity to the masses by erasing a marker that generated difference. That was the mechanism through which the reforms in 1921 and later in 1945 favored "civic" nationalism. But we should reflect for a moment here again on the limitations of using the categories civic and ethnic. The separation that they presume is about two ways of engaging with the national polity. But there are no clean cases and tidy distinctions. A careful analysis of any polity reveals mixtures or, better yet, different goals and agendas of nationalist agitation and politics that are variably inclusive and exclusionary depending on the historical context. As there are no predetermined paths, the analysis of nationalism responds better to historical inquiries than to sociological modeling.

The orthographic question, when scrutinized beyond the *evènement*, allows one to glimpse its dynamic within the *conjoncture*. The fact that it parallels the political history of the modern Bulgarian state from 1878 permits the linkage of nationalism to the long-term political development of the country. In other words, the picture of nationalism that it elicits is one that intimately grapples with one of the most central political processes of the modern Bulgarian state—the question of democracy and popular sovereignty. As stated above, there is good reason to link the three revolutionary reconceptualizations of the Bulgarian polity—independence, the First and Second World Wars—to the simultaneously occurring crucial and radical interventions into the ethno-linguistic nature of the nation via orthographic reform. That Stamboliiski's agrarian regime and the communist regime both adopted a similar stance is consistent with the central tenet of these social experiments, that of popularization and democratization within the nation. Rhetoric aside, especially that of the outwardly non-national communist ideology, the orthographic reform reimagined the national community in Bulgaria. While the critique of the civic/ethnic antinomy from within necessitated the demonstration of shifts using the problematic binaries of the model, the contextual approach that this chapter advocates could further deconstruct the reimagination of the national community. Even as the "civic" base of the nation was broadened with the orthographic reform, the consolidation of the national language contributed to a new set of exclusions. Attention to the changing hierarchies within the nation can reveal the constellation of contradictions it tried to cover.

The contest for the soul of the nation and the different potentialities and outcomes could be very succinctly summed up in a question that owes its terminology to Benedict Anderson: Whose imagined community? This question plumbs to the depths of what is at stake in imagining the nation as a sociocultural project. This question, however, is not mine and has been raised with slightly different connotations by Partha Chatterjee's homonymous article from 1991 that became the first chapter of his *The Nation and Its Fragments* in 1993.[63] Despite the differences which will become apparent below, the two projects—mine, as developed in this chapter, and his—converge in their implications. In order to situate this relationship and to establish the limits to the points of contact, a few words are necessary about the work of the Subaltern Studies Collective and the vast intellectual impact it has made in the past decades.

Subaltern studies, beginning with the publication of its volumes from the early 1980s, began as a project to problematize Indian and Southeast Asian history in its particular relation to the colonial and postcolonial condition. The impact of this reevaluation by the multidisciplinary scholarship of the collective was profound precisely because the frameworks within which the particular studies were situated had implications for European and world history as well. For example, the critique of the liberal historicist mythology and the exposure of the gap between its promise and realization that Dipesh Chakrabarty performed in *Provincializing Europe* reverberates in Marxist theoretical frameworks about the function and condition of modernity, particularly when these conceptions are forced to account for difference.[64] Such relational critiques, when connected internally to the subaltern position of groups within society, such as the peasantry, opened new avenues for deconstructing the hegemonic practices imbedded within discourses of the "universal" in relation to the previously subsumed and marginalized particular.

The subaltern studies project is not static and has undergone changes in its thematic focus and editorial leadership. The widening of its claims from its humble Marxist origins under Ranajit Guha is mirrored in the critiques of it that strive to keep up with the widening of its influence. Thus, while Sumit Sarkar could draw attention to the relative marginalization of the micro-historical work on the subaltern subject in "The Decline of the Subaltern in *Subaltern Studies*" in 1997, the most recent salvo from Vivek Chibber's *Postcolonial Theory and the Specter of Capital* is revelatory of the status that subaltern studies has achieved when it is used by him as a theoretically cohesive proxy for the variability of postcolonial theory.[65] The debate between Chibber and Partha Chatterjee at the Historical Materialism Conference in New York on April 28, 2013, entitled "Marxism & the Legacy of Subaltern Studies," turned out to be far from the swan song for a spent movement as was forecast. Although the parameters of the debate restricted the discussion to only a few topics, nonetheless the central divergence over the treatment of difference is emblematic. Below is a transcription of the conclusion of Partha Chatterjee's intervention; it functions as an iteration of a manifesto that stakes out claims, defends choices taken, and through *praxis* links theory to politics:

> One can only acknowledge that the debate between universalism and its critics continues and will not be resolved in a hurry. The choice between the two sides at this time is indeed political. The greatest strength of the universalist position is the assurance it provides of predictability and control over uncertain outcomes.

It is the assurance that past history is a reliable guide to the future, which leads Chibber to insist that capitalism or the struggles of subaltern classes must be the same everywhere. . . . The critics of universalism argue that the outcomes are unknown, indeterminate, and hence unpredictable. They accept the challenge of risky political choices based on provisional, contingent, and corrigible historical knowledge. Since its early work, which Chibber has dissected in his book, Subaltern Studies went in several new directions, especially after the important interventions by Gayatri Spivak and Gyan Prakash. Their historical practice does not rule out the rise of new universalist principles, but these, they insist, must be forged anew. What is certain is that the working classes of Europe and North America and their ideologues can no longer act as the designated avant-garde in the struggles of subaltern classes in other parts of the world. Historians of Subaltern Studies have only attempted to interpret a small part of these struggles, and changing the world, needless to say, is a job that cannot be entrusted to historians.[66]

What is left unsaid in this quotation, but which is implicit, is the function of the subaltern to the project. It is through paying attention to the presumably powerless, malleable, or marginalized that a counterexample to the hegemonic narrative can be produced, the site of its failure or contradiction can be pinpointed, and a new conceptualization can be produced.

This critical thrust of subaltern studies is what makes it such a useful theoretical frame in relation to this monograph on political agrarianism in general and to its treatment of nationalism in particular. This relation is much more than a simple parallelism through the attention to the peasantry as a subaltern subject. The deeper claim is that through the study of the subaltern subject, in this case, the peasant, certain important critiques can be made that otherwise are obscured. In the case of subaltern studies as it started in the 1980s, the central issue was how to continue to apply Marxist theory in empirical situations that did not conform to expectations and challenged orthodoxies. In effect, it was an internal critique that argued that the exceptional case from the "periphery" is of paramount importance to the nuancing and clarifying of capitalist practice. Of course, the reverberations of this critical analysis were imbedded within the colonial and postcolonial history of India and Southeast Asia, and so the divergences from the liberal promises of the nation-state produced a similar shake-up in the other pillar of modernity: nationalism.

When Partha Chatterjee wrote *Whose Imagined Community*, he performed a series of such critical moves with regard to nationalism in India. He challenged Eurocentrism by arguing that the Indian case conforms to modular forms imported from Europe neither quantitatively nor qualitatively, because the implication is that the non-European imagination "must remain forever colonized."[67] Further, he profited from the insight that the circularity of standard nationalist history, which could also be considered as an elite narrative, or one that overemphasizes nationalism as a political movement, is bound to reproduce itself: "In fact, since it seeks to replicate in its own history the history of the modern state in Europe, nationalism's self-representation will inevitably corroborate Anderson's decoding of the nationalist myth."[68] Based on this, Chatterjee locates and describes a prior location of sovereignty for nationalism in the inner spiritual domain of the social imagination that is confronted by Western material superiority. While the case of Bengal that he fleshes out is not transferable either in its particulars or in its more generalized structure to the rest of the world, the very fact that he locates it opens up the possibility of alternative readings in other arenas. This, in my view, is the inspirational function of the subaltern studies critique.

More particular applications are problematic, and I would like to flesh out this point a bit as an explanation of the distance I maintain from subaltern studies, even as I continue to demonstrate how it informs this work. Said differently, I would like to tackle here the temptation in Eastern European historiography to tie itself to broader theoretical movements in order to escape the parochialism of area studies, but often at the cost of problematic transplantation. This situation is not unique to Eastern European studies. Daniel Clayton's conclusion to his review of Chakrabarty's *Habitations of Modernity: Essays in the Wake of Subaltern Studies* grapples with the temptation of transferability: "Chakrabarty gives us some important insights into how and why modern Indian history has become the locus classicus for so much theoretical production in the sprawling field of postcolonial studies. By the same token, he causes us (or at least this reader) to ask: how might we think and write about forms or habitations of postcoloniality that deviate from canonical Indian intellectual conceptions of this term and condition."[69]

One of the best engagements with this problem that emerged from within Eastern European studies was the provocative article "Thinking between the Posts: Postcolonialism, Postsocialism, and Ethnography after the Cold War" by Katherine Verdery and Sharad Chari.[70] The article's goal was to suggest points of contact between postcolonial and post-socialist studies. The three concrete areas where the authors think that this can best occur are in work on imperialism, the jettisoning of the Cold War division of the world into Second and Third Worlds, and attention to the biopolitics of "state racisms," in their terms, but which actually devolves to the treatment of the "other" in categories such as "internal enemies." Although stimulating to consider, the actual content at these sites is debatable. Maria Todorova, for example, problematizes the characterization of state relations in the Soviet Bloc as colonial, even as she herself considers the limits of the applicability of postcolonialism, partly in relation to her earlier distinctions between Balkanism and Orientalism in her "Balkanism and Postcolonialism, or On the Beauty of the Airplane View."[71] I would argue that the content of this debate is as unimportant as the determination of sites of contact so long as critical inquiry does not contravene the production of meticulous thick description. To this reader, the most fruitful contribution of Verdery and Chari's piece is contained in the critical possibilities imbedded in the rescuing of the term "postsocialism" from the anti-communism of the transition.[72] "Over time, 'postsocialism' too came to signify a critical standpoint, in several senses: critical of the socialist past and of possible socialist futures; critical of the present as neoliberal verities about transition, markets, and democracy were being imposed upon former socialist spaces; and critical of the possibilities for knowledge as shaped by Cold War institutions."[73] I hope that this present work serves as a fruitful illustration of how the use of the subaltern project as inspiration can produce a nuanced conversation between Eastern European studies and postcolonial theory.

If the legacy of subaltern studies is this spirit of critical inquiry that opens space for reevaluation, then it is especially important to avoid tendencies toward ossification. In *Whose Imagined Community?*, Chatterjee provides a genealogy of the impetus toward reevaluation by referring to the "'area specialists,' the historians of the colonial world, working their way cheerlessly through musty files of administrative reports and official correspondence," out of whose ranks a figure like Benedict Anderson emerged.[74] The humility of painstaking work that can identify inconsistencies and difference is in fact the method by which subaltern studies has been able to supersede "elitist" historiography or to provincialize Europe.

However, there is a troubling tendency to reify Europe even as the universalist claims of Europe are undone. It is surprising that Chatterjee reiterated a point in his debate with Chibber in 2013 that he had codified back in 2001 in his synthesis "Subaltern History" for the *International Encyclopedia of the Social and Behavioral Sciences*:[75]

> The historical problem confronted by Subaltern Studies is not intrinsically a difference between West and East, as Chibber repeatedly insists. The geographical

distinction is merely the spatial label for a historical difference. The difference is indicated, let me insist emphatically, by the disappearance of the peasantry in Capitalist Europe and the continued reproduction to this day of a peasantry under the rule of Capital in the countries of Asia, Africa, and Latin America.... That is also why, despite the apparent similarity, Subaltern Studies could never have been carried out in the same way as history from below in Europe. The recounting in the latter body of work of the struggles of peasants and artisans in the period of the ascendancy of capitalism in Europe was inevitably written as tragedy since the ultimate dissolution of these classes was already scripted into their history. Should we assume the same trajectory for agrarian societies in other parts of the world? Does a different sequencing of capitalist modernity there not mean that the historical outcomes in terms of economic formations, political institutions of cultural practices might be quite different from those we see in the West?[76]

The contradiction in this claim is that while it justifies the provincialization of Europe from without, it denies its possibility from within (in this case via a Thompsonian history from below). Subaltern studies opens up space for the other—in general India and Southeast Asia—in parallel to Europe. That is done by demonstrating that neither is there a simple transplantation and accompanying imposition of ideas or institutions nor are all other nationalisms or modernities derivative. Because of the "West versus the Rest" duality, the argument for difference, multiple modernities, unfortunately is not brought back to Europe. Europe remains in some important respects homogeneous. I would claim that that critical logic must be and can be brought back to Europe to perform the rupture with elite historiography. Otherwise, it would be an intolerable irony for subaltern studies to be the expression of European elite historiography.

An example would more clearly illustrate my point here. In the chapter "The Nation and its Peasants," Chatterjee begins his analysis with a portrayal of the fate of the European peasantry. Two claims cover the range of European experience. "In Western Europe, the institutionalization of a modern regime of power coincides with or follows a process of the extinction of the peasantry. Even in France, where it survived as a sufficiently large mass of the population in the second half of the nineteenth century, the peasantry was associated with such supposedly aberrant political phenomena as Bonapartism and had to be systematically disciplined and transformed into 'Frenchmen.'"[77] He makes the following claim with regard to the other half of the continent: "Further east, the peasantry figured for more than half a century as a hub of a fierce debate between populists and Marxists over its role in a revolutionary Russia.... In the end, the matter was settled in Russia by the elimination of the peasantry under the collectivization program of the 1930s."[78] Nowhere is the European peasantry accorded any agency; Hegel speaks about it. The problem of the veracity of the truth claim about Europe disappears in the logical structure of Chatterjee's deduction: Europe and its relationship to its peasantry is A, the colonial East and its relationship to the peasantry is the opposite of A (extinction vs. survival) and hence the conclusion about the implication of historical difference and the necessity for autonomous analysis. The logical deduction is valid irrespective of the truth value of the premises, but it cannot be sound, if either premise is invalid.

Even as I argue that this representation of Europe by Chatterjee is in an unsatisfactory reduction that is false as far as interwar agrarian Europe[79] is concerned, I want to stress that essentializing Europe is not necessary for provincializing it. The subaltern critique is consistent with a similar project that achieves that provincialization from within. Therefore, enriching the European experience with the complex history of the alternative that the agrarian movements of Bulgaria, Czechoslovakia, and Yugoslavia elaborated in no way invalidates the enrichment brought about by the accommodation of the colonial East. Tongue in cheek, I cannot help but remind the reader that the peasant alternative was not eliminated in interwar Europe by capitalism. Nor is the very complex path of the peasantry's disappearance in the socialist period reducible to the typology of violent Stalinist war against the peasantry.[80] Some points of departure for how the agrarian alternative was brought to an end in the years following the Second World War will be the topic of the last chapter, but suffice it to say here that this is why its history in this work *cannot* be and *is not* written as tragedy.

The subaltern inspiration can be turned against its own essentializations through the insight that understandings of Europe's interwar period have been impoverished by the absence of the peasantry as a subject, a condition which is doubly ironic since it treats a moment when the peasantry sought to shrug off its subaltern condition and projected itself as a transformative modern political subject. And while this thread runs through the entire study of the golden age of the European peasantry, in this particular chapter, the inspiration of subaltern studies is activated at the level of representations of Eastern European nationalism, where I argue that close attention to the relationship between agrarian politics, nationalism, and the reconfiguration of the nation reveals processes that are otherwise missed and through their inclusion can do nothing but force an enrichment of the static models which unfortunately dominate scholarship even to date.

This chapter has been a demonstrative experiment in two directions. The first pertains to agrarianism itself and has sought to demonstrate the horizon of possibility that motivated the golden age of agrarianism: in relation to nationalism, it showed the political ethics beneath a topic that has attracted little attention but which, as I have demonstrated, is pregnant with significance. The second pertains to the significance of agrarianism for European history and the social sciences and has sought to demonstrate that a neglected topic can offer a unique point of reference from which to interrogate the theoretical and empirical foundations of other subfields and disciplines. Both directions celebrate the exciting potential still available to research into agrarianism and in a small way mimic the excitement that agrarians could harness to their cause in the interwar period as they sought to imagine a more equitable modernity. The next chapter's purpose is to qualify this potential along the line of Marx's dictum that men make their own history but do not make it as they please. The chapter illustrates the tension between aspirations and limitations that hindered the expression of the agrarian potential.

# 4

# Between Aspirations and Limitations

Raiko Daskalov, Stamboliiski's partner in the Radomir Rebellion, had been sent to Prague in May 1923 as the minister plenipotentiary to Czechoslovakia. On June 9, 1923, he was in Prague and there he began to organize the response to the coup in Sofia. He founded the Representation of the Bulgarian Agrarian National Union Abroad. On August 26, 1923, an agent of IMRO, Iordan Tsitsonkov, carried out the death sentence that had been passed by IMRO against Daskalov and that had failed once before in December 1922. Todor Alexandrov, one of the leading figures in IMRO, sent out a circular to the IMRO *vojvodas* on August 27, 1923: "On the 26 of August three revolutionary bullets brought down in Prague the former Minister in the Cabinet of Stamboliiski, Raiko Daskalov. Having covered his tracks for a long time, having almost gone completely underground for about three months, the killed could not hide from the watchful eye of IMRO: she found him and for his crimes against Macedonia and her holy cause, she punished him with death."[1] Even though Tsitsonkov was captured at the site of the assassination, he was exonerated at his trial. Outrage from the BANU émigrés and the Czechoslovak public, and even diplomatic pressure from the Kingdom of Serbs, Croats, and Slovenes (not because there was a special affection for Daskalov but out of its struggle against the IMRO) forced a second trial at which Tsitsonkov was sentenced to twenty years in prison.

After the assassination of Daskalov, the Representation of the Bulgarian Agrarian National Union Abroad was headed by Aleksandur Obbov, the deposed Minister of Agriculture and State Property. Under Obbov, the official purpose of the Representation Abroad was (1) to organize European public opinion against the regime in Bulgaria, (2) to save the comrades in danger inside Bulgaria, (3) to inform the public about events in Bulgaria by publishing *Zemledelsko Zname* and contributing to the *Bulletin of the MAB*, and (4) to organize the (Bulgarian) youth abroad.[2]

It is tempting to concede the point to critics that present the Representation Abroad as an information center. The initial efforts to form a united front with the communists in 1924 failed and thereafter Obbov and Kosta Todorov engaged in propaganda to defend themselves from their representations as communist agents by the Tsankov and then the Liapchev governments, but also the ones that emerged from within the BANU as the organization started to fracture. The anti-communist component grew also in proportion to the threat of communist infiltration as well as communist advances in the countryside due to the BANU's weakness.

The Representation Abroad survived until the partial Amnesty Laws of 1932 and 1933 on a meager initial sum of two million leva, funds it could collect from the gardener associations, members of the BANU, and, crucially, financial and logistical aid from the Czechoslovak Republican Party. In some ways, the support in Czechoslovakia was decisive to the survival of this organization, but in others, it kept it barely above the subsistence level. For example, the Czechoslovak foreign ministry rejected the demand by its Bulgarian counterpart on August 8, 1923, to extradite Obbov over specious accusations that he had stolen BANU property.[3] This defense extended to the rest of the Bulgarian agrarian cadres, even though the links with Kosta Todorov in Belgrade to explore the possibility of armed resistance brought complaints from Bulgaria that Czechoslovakia's continued succor of these "criminals" was compromising the relations between the two countries.

The Representation Abroad received office space from the Czechoslovak Republican Party and used their printing facilities for *Zemledelsko Zname*, which began to reappear on September 15, 1923, after it had been closed in Bulgaria. The Bulgarian émigrés for a while received stipends and some had subventions to study at the Cooperative Institute. While this assistance was crucial, Czechoslovak policy was measured. Leftist revolutionary tremors had the potential for embarrassment, and so Obbov was progressively marginalized. This was aided by the fact that as the BANU splintered, the émigré leadership too engaged in personal attacks and infighting among itself. At the start of the Republican Party Congress on September 5, 1925, neither Nedialko Atanasov, the former Minister of Rails, Posts, and Telegraphs, who had spent almost a year in prison after the coup but had emigrated after his release, nor Obbov were allowed to make a statement. Obbov had called Atanasov a Bolshevik spy and the latter was planning to complain and turn the tables at the congress.[4]

This factionalism made the Czechoslovak position which favored legal activity and a more centrist approach easier when it came to promoting individuals or lending support. Officially, the position was to unite the BANU. Yet, in 1932, a year after the Democratic Alliance lost the elections in Bulgaria to the People's Bloc that included the BANU's Vrabcha 1 of Dimitur Gichev, Milan Hodža summoned the Bulgarian minister plenipotentiary in Prague, Pancho Dorev, to protest his portrayal as being "the financial minister of the Bulgarian émigrés" in Liapchev's paper, *Demokraticheski Sgovor*. Hodža stated, "To Obbov, Atanasov, and others we gave a few thousand crowns monthly, but on the condition to not engage in any active politics. As soon as Obbov transgressed that condition, we stopped being interested in him and he left for Paris with not one sou. . . . I am not with Obbov and his 'Pladne' . . . I and the party are for the active and positive politics of all agrarian parties, for cross-party cooperation in support of democracy and the constitution." In fact, Hodža supported the coalitional manner in which BANU's Vrabcha 1 had come to power, thus emulating the example of the Republican Party and further stated, "This is your strength in comparison to your neighbors."[5] The hopes for coalition were dashed with the May 19, 1934, coup by Zveno which abolished all political parties.

As stated above, the constraints on the Representation Abroad make it an easy target for dismissal as an information center. On the contrary, however, I would argue that these limitations notwithstanding, the Representation Abroad's greatest significance was that it could preserve the critical and to some degree independent position of Bulgarian agrarianism. It was spared the initial terror and the subsequent harassment that plagued the BANU in Bulgaria and resulted in its fractioning. Its biggest achievement was that at a time of seeming hopelessness that anything could be done to restore political normalcy and a semblance of democracy in Bulgaria, its members tenaciously kept the spark alive for themselves, the beleaguered BANU, their agrarian partners in the MAB, and in front of European public opinion.

This would not amount to that much, if it only concerned a few dozen already politicized grizzled veterans that could not accept defeat even in the context of exile. The difference between these men and those colleagues of theirs who formed the International Peasant Union after the Second World War (see Chapter 6) is that agrarianism was still alive as an idea and still carried a potential that could inspire its base. To illustrate this, I would like to bring in an extremely rare instance in which the documentary record reflects a radical intervention from the base, an instance that allows a momentary engagement with history from below.

On May 27, 1927, responding to meetings of Bulgarian émigrés in Paris, Belgrade, and Prague, Obbov and Kosta Todorov attempted to exert some control over the discussion by channeling it in an "Appeal to the Bulgarian United (*zdruzheni*)[6] Agrarians Abroad." In the appeal, they summarize the content of these meetings as revolving around the following two questions: "Do we not lose too much ideologically, situated as we are abroad, as we live ideologically divided far from our motherland, if we do not follow and participate in the ideological development of our organization?" and "What do we give [now], divided as we are, and what could we contribute to the cause of the Bulgarian peasant, if we could live organizationally united?"[7] The Representation Abroad packaged the concerns from the émigré meetings in these questions in such a way so as to be able to maintain its leadership and assert some control over the way dissatisfaction was being expressed. Thus, to recapture the leading role, their appeal called for the creation of a Cultural-Educational Union of the Members of BANU Abroad.

From the documentary evidence regarding this organizational effort that departed from the center, the Representation Abroad, everything appeared under control. On June 20, Stancho Trifonov, writing in the name of Obbov, wrote to the comrades in Vienna, Brno, Munich, and Stuttgart to inform them that the center was in agreement with their arguments to postpone the founding congress of the new organization until after the end of the summer holidays, so that the students who were to spend the summer break in Bulgaria could return.[8] What the letter did not mention, however, was the hard-to-contain excitement and mobilization that was bubbling from below.

I transcribe below a significant portion of one such letter. Even though it forms a part of the story of the organization that was to eventually culminate in the creation of the Union of the Bulgarian United (*zdruzheni*) Agrarians Abroad, for me its greatest significance is as a testament to the stake individuals had in political agrarianism. Its expression is made all the more poignant by the context of 1927, when Andrei Liapchev, who had taken over from Tsankov as the head of the Democratic Alliance, was consolidating its rule which was to last until 1931.

On June 5, 1927, Peicho Pop-Petrov wrote a letter addressed to the temporary committee of the Cultural-Educational Union:

> I received the appeal and the draft statutes and I thank you very much. The initiative is extremely timely and worthy of praise. Up to now I have always seen an emptiness in the life of the BANU, namely the lack of a journal which could serve as the ideological platform of the organization. Much more fundamental is the need of those who are scattered abroad and thus are denied the possibility to participate firsthand in the life of the Union. It is gratifying that today, the cream of the emigration is undertaking a task, which although difficult, will

produce good results. The four-year-long road that the first two members of the temporary committee have behind them and which not only they, but also their ideological comrades can regard with pride, is sufficient guarantee for success. These four difficult years for BANU and [four years] of difficult hardship for the emigration testify to the constancy, energy, and fighting spirit of the leaders, of which we, the rank and file, can only be proud. . . .

I am and will remain a partisan of the ideas for which the Union is crucified, and I am ready to give my support for its triumph. I have not been a member of BANU because then I was still young and this is the fifth year since I left Bulgaria. I was a member of the Youth Agrarian Union in Bulgaria and I hope that I will be accepted without reservations in the village *druzhba* even though my future profession is foreign to agriculture. A person must have love for it [the village] and must work for the cultural advancement of the estate from which he comes, to be a member of BANU.

In Stuttgart, I am the sole one of ours [people, agrarians]. I would be glad to join the group in Munich, Karlsruhe, or another. If the journal appears, please send an issue to me as well.

With many wishes for success I send you my brotherly greetings.[9]

Peicho Pop-Petrov was to some extent an exception. His peers in Brno were more cautious: "Given the extraordinary circumstances in Bulgaria the founding of this Union should not be rushed; students are still afraid to enter into such a Union in order not to be suspected of having links to émigrés, and upon their return in Bulgaria [because of this] to be mistreated by the government and irresponsible forces there."[10] Nonetheless peers with his convictions gathered at the founding congress of this Union in Paris between April 11 and 13, 1928, and began a revolt from below. In meetings on November 11 and 12, 1928, the Parisian section of the Union of the Bulgarian United (*zdruzheni*) Agrarians Abroad forcefully articulated its dissatisfaction with the course of the BANU and the factionalism among its leadership. Its resolution on the state of affairs in BANU is a forceful indictment that puts to rest the thesis of a disconnect between a manipulative leadership and a passive peasantry:

## RESOLUTION about the situation of BANU

The unified student agrarians in Paris, at their meetings on 11 and 12 November, discussed the debates and struggles within BANU caused by the latest embroilments, and found:

1. that the internal struggles within the Union are only among the leaders of the Union, whereas the peasant masses are united in solidarity and true

to the main principles of the agrarian movement, such as they had been pronounced by our late teacher and leader Al. Stamboliiski.
2. that these struggles among the leaders, however, create despair and disgust amid the peasant masses, and are considered the result of personal ambitions, without any ideological or tactical reasons.
3. that in fact until today the main representatives of these groups have not communicated either their programmatic, nor their principal or even tactical differences, to justify this excessive hostility among themselves.
4. that, on the contrary, the heavy accusations are based solely on the errors of separate individuals and not on the ideas that these individuals promote and defend.
5. that this situation causes a decline in the powers of the Union and hinders it in playing the role that it should have, because of its size and the significance of its great principles, in the political development of the country.
6. that the government and the reactionary forces want to use this situation in the Union, by nourishing these struggles with all available means, either through promises that make some of the leaders believe in the possibility of political combinations, which secure them ministerial posts, or through threats of persecution and terror.

    All these dark forces have gathered to support these struggles in the Union and to corrupt their participants, to kill the morality and the faith in victory of the peasant masses, and to diminish the power of their opposition to the present regime in power.
7. that the communists, on the other hand, having lost all hope to restore their party, are looking for revenge, attempting to influence and take over the leadership of the Union, using the same methods and means to influence separate individuals, advocating a "unified front."

## DECIDED

1. It rebukes these struggles as ones that paralyze the forces of the Union and hamper it from taking its proper place in our political life.
2. It invites the fighting parties to clearly state the ideological or tactical differences that divide them.
3. It appeals to all unified agrarians, who are not divided because their shared interests order them to move together, not to succumb to the passions that feed these struggles and lead to cleavages in the Union.
4. It finds that the only exit out of this situation are the congresses, in which the Statute of the Union will be respected agrarians will be correctly

represented, and the elected governing bodies will be based on the trust of the unified agrarians.

For the meeting:<br>
Chairman: DIMIROV<br>
Paris, 12 November 1928[11]

The creation of the Union of the Bulgarian United (*zdruzheni*) Agrarians Abroad was by no means a momentary blip. It began the publication of the journal *Zemledelets* from May 1928 and issued it for two years until the publication had to cease due to lack of funding. The first issue of 1929 reported on the Union's attendance of the founding congress in Paris of the *Parti agraire et paysan français*, on January 26 and 27, 1929. Karel Mečíř, the secretary general of the MAB attended the congress alongside Ferdinand Klindera, the president of the Prague *Centrokooperativ*. The Bulgarian student émigrés had the following to say of this event and the significance of the BANU's agrarianism:

> The founding of an agrarian party in France is a slap to those Bulgarian politicians [*politikani*], which even in clearly agrarian Bulgaria deny the right of life to BANU. The Bulgarian agrarians cannot but rejoice at this significant event and see that the road that they chose 30 years ago was the most correct and that the theories of agrarian organization [*zdruzhavane*- based on the *druzhbi*- (the organizational units of the BANU at the local level)] and estate struggle [BANU corporatism] are not "vulgar," as our "learned" Bulgarian opponents call them every day, but are [of] such [quality] that today, one after the other, the cultured peasants of Western Europe copy them word for word.[12]

The ironic inversion of hierarchies in this quotation deserves some commentary and contextualization. The last issue of the *Bulletin of the MAB* of 1925 included the following celebratory editorial note when it reprinted a letter by Gabriel Fleurant-Agricola addressed to Karel Mečíř and which was entitled "Rural Democracies."

> Mr. G. Fleurant-Agricola, one of the most eminent propagators of the peasant idea and a fervent herald of rural democracy sent us this article. We publish it with a joy that is all the greater in that it proves how thinkers that base themselves on an agrarian philosophy, always and necessarily arrive to the same conclusions. One can say that the ideas of Mr. Fleurant are practically the same as those we strive to propagate in Czechoslovakia in our campaigns. And our eminent collaborator in France would without doubt be happy to hear that the latest developments at home have so amply justified the accuracy of the theses advocated by him.[13]

While the first letter dealt generally with the amelioration of the condition of the countryside, the second letter which subsequently arrived and was published in the first issue of 1926 argued in favor of a nonpolitical, strictly professional peasant internationalism predicated on conservative organic work among the peasantry. This letter prompted a polite, but firm rebuttal by Mečíř that insisted on the indispensability of peasant political organizations.[14]

Gabriel Fleurant-Agricola would eventually accept the vision of a politically organized peasantry, and he was the founder of the *Parti agraire et paysan français*, whose congress the Bulgarian student émigrés attended in 1929. The French Agrarian and Peasant Party held its constituent assembly on March 23, 1928, and joined the Green International on May 2, 1928.[15]

Eugen Weber's *Peasants into Frenchmen* engendered numerous critiques to its modernization theory transformation of peasant consciousness.[16] Irrespective of one's take on this debate, there is obvious fundamental consensus on nationalization and politicization in the French countryside in the interwar period, one that demographically only tipped in favor of the urban in 1931. Less well known is Robert Paxton's *French Peasant Fascism: Henry Dorgeres' Greenshirts and the Crises of French Agriculture, 1929-1939* that engages with the Right-most expression of peasant politics in the interwar period. I could jokingly paraphrase the title into *Peasant Frenchmen into Fascists*, but that would do a disservice to the meticulous excavation of the politicized peasant dissatisfaction with the economic, cultural, and political crises confronting the French peasantry around the time of the Great Depression and after.

More work on the various and complex paths to politicization of the French peasantry in the interwar period is welcome beyond its most provocative manifestation. After all, Gabriel Fleurent's centrist *Parti agraire et paysan français* became sufficiently disillusioned with the politics of the Third Republic to join Dorgeres's *Comités de défense paysanne* in the formation of the *Front Paysan* in 1934. The third component of that front was the *Union nationale des syndicats agricoles* headed by Jacques Le Roy Ladurie, the father of the historian of the peasantry, Emmanuel Le Roy Ladurie, whose trajectory led him to become Minister of Agriculture in Vichy but who would also join the resistance by 1943.

The elaboration of this complex history, entangled as it is in European agrarianism, is an important endeavor. Granted, this can be achieved by inclusive, multinational, scholarly initiatives initiated in the center. One such example occurred in the 1960s that incorporated Polish, Romanian, and Russian scholarship and which Henri Mendras recounts in "The Invention of the Peasantry: A Moment in the History of Post-World War II French Sociology."[17]

But there is space to energize the evaluation and excavation of this area of French history through engagement and dialogue with the historiography of the European periphery.

The letter of Pop-Petrov, the 1928 resolution, and the perspective on French agrarianism quoted above signify for me the integral part of agrarianism that made it an inalienable part of the interwar period. Above all else, agrarianism was the expression but also the constitution of a moral economy in the peasant that altered his relation to modernity from a transitory object to an agent with a stake in its constitution. For reasons which will be explained below, this core has been entirely neglected by scholarship that has sought to synthesize agrarianism above the level of national politics. That in itself is a question of interest because even if there are significant differences in the degree to which the national histories of the agrarian parties have been elaborated, they nevertheless have succeeded in producing a nuanced reading that reflects the significance of the agrarian moment. The failure to properly situate agrarianism in regional surveys is easier to overlook because that can be remedied more easily. The more serious matter concerns the shortcomings of the more theoretically minded scholarship that has tackled the problem of agrarianism.

To begin, a fundamental problem is caused by an excessive focus on a limited interpretation of ideology. A quotation from G. M. Dimitrov's chapter on Agrarianism in Feliks Gross's *European Ideologies: A Survey of 20th Century Political Ideas* appears ubiquitously without either its introductory clause or the sentence that follows, or without any context whatsoever: "Agrarianism does not yet possess a systematic doctrine of fundamental principles or a coherent philosophical structure of values." The whole statement reads: "Naturally, being recent, *agrarianism does not yet possess a systematic doctrine of fundamental principles or a coherent philosophical structure of values.* Nevertheless, as may be seen even from this brief account, its ideological and theoretical argumentation tends to rely on the dynamic scientific attainments of the age taken as a whole [italics mine]."[18] The italicized passage is taken by authors as evidence that even for one of the agrarian leaders, agrarianism was nebulous. The most recent example is *Agrarismus und Agrareliten in Ostmitteleuropa*, a work which will shortly be analyzed below.[19] The uncontextualized sentence fails to acknowledge that a few paragraphs earlier, G. M. Dimitrov, while acknowledging that agrarianism was not entirely fleshed out, still opens his "brief account" with the sentence: "In its fundamental principles, Agrarianism tends to be an ideology of political and economic democracy based on the idea of cooperative syndicalism."

The obsession with ideology is the bane of efforts to theoretically grapple with agrarianism. The struggle by scholarship to come up with a cohesive representation circularly also produces one of the major reasons that are given for agrarianism's historic failure—that is, the absence or lack of a cohesive ideology. This denigration comes not only from the competing ideological currents—from the communist Left, the liberal Center, or from the Right. It also intimately tied to the overly ideologized mode of scholarship that characterized the study of communism, the Soviet Union, and after the Second World War, Eastern Europe. What can be seen in the reductive logic of the totalitarian thesis can be extended to this case. It is the obsession with ideology that is the defining characteristic, the determining thread that runs through history, through analysis, and permits moral judgment. It is a simple equation: characterize, classify, critique, and overcome.

The focus on ideology misses the part of agrarianism as radical politics, that is not about administration, but a contest over the distribution of power; in the words of Francisco Panizza, "Politics is about challenging the institutional order with the radical language of the excluded, but it is also a dimension of the practices that make institutions operative, and contribute to both their existence and erosion over time. As such, it operates in the spaces between the political logic of the permanent revolution and the technocratic logic of the end of history."[20]

As radical politics par excellence, for which people dedicated and sacrificed their lives, agrarianism occupied and elaborated the tertiary space between the two other dominant modes of modernity, while holding onto the platforms that were elaborated around the First World War, all throughout the interwar period: in the Bulgarian case written by Stamboliiski in prison, for the Croat Peasant Party in 1921 with an addendum in 1922, for the Czechoslovak Republican Party, in 1922. Agrarianism in action was thus not a logical elaboration of some fundamental principles but contextual praxis that responded and adjusted to its national, contextual specificities as well as to the brutal environment of the interwar decades.

It is not surprising therefore that there was a multiplicity of maneuvers and directions that agrarianism took. This work strategically examines the different manifestations in three countries in order to show three vertices of the elaboration of agrarianism: the radical moment in Bulgaria, the national direction in Yugoslavia, and the centrist position in Czechoslovakia. That which is worthy of analysis, and which heretofore had not been satisfactorily attempted, is to explain what held these differing movements together in a matrix and network of information and cooperation. The answer this work proposes is imbedded within its definition of agrarianism as a third road between capitalism and communism. It was the awareness of the agrarians that they had found their place on the world stage—not in one country, or two, but regionally, and in their hopes, in Europe and the world—that produced initiatives such as the Green International.

To map the scholarly terrain that I have been mentioning, I would like to give two examples of approaches to agrarianism. The first is the approach to agrarianism that adopted the lens of populism. In Ernest Gellner's and Ghita Ionesco's *Populism: Its Meaning and National Characteristics*, this was accomplished in the classificatory mode.[21] Even though the work begins with a riff from Marx to establish its serious ambitions—"A spectre is haunting the world—populism"—it is a product of modernization theory, and implicit in it is a model to explain backwardness. Ionescu, who wrote the chapter on Eastern Europe, asks: "This of course leads to a much vaster question, whether the peasantry as a dominant class is historically able to effect the revolution of modernization."[22] Instead of answering himself, Ionescu quotes Barrington Moore from the *Social Origins of Dictatorship and Democracy*: "The peasants have provided the dynamite to bring down the old building. To the subsequent work of reconstruction they have brought nothing; instead they have been its first victims."[23] Peter Wiles then provides the operating schema for the volume. In the chapter entitled "A Syndrome, not a Doctrine: Some Elementary Theses on Populism," the author lists twenty-four points which, à la Brzezinski and Friedrich's characteristics of totalitarianism (they only had six), map out the phenomenon. Points four through six are gems: "(4) Populism is in each case loosely organized and ill-disciplined: a movement rather than a party, (5) Its ideology is loose, and attempts to define it exactly arose derision and hostility, (6) Populism is anti-intellectual. Even its intellectuals try to be anti-intellectual."[24]

The *Agrarismus in Ostmitteleuropa* project at the European University Viadrina brought together a remarkable collection of scholars from the region in a series of conferences from 2007 to 2010. I attended the one in 2009, and even though the individual contributions had merit as particular scholarship, the laudable effort to reexamine agrarianism was hampered by an insistence on a structural synthesis. For example, I was particularly troubled by the splitting of the region into Central Europe and the Balkans through the rubrics of "aristocratic agrarianism" and "peasant agrarianism." When the final report was published in 2010, its summary began with, "As initial hypothesis, East Central European societies entered the modernity not only with the burden of peripheral structure, and the heritage of long lasting foreign rule, but as incomplete societies too."25 How do all these structural "preconditions" enlighten the examination of agrarianism as a political movement and as subjectivity, other than to "other" an already liminal subject and impose a limiting framework? The latest product from this project is *Agrarismus und Agrareliten in Ostmitteleuropa* which was mentioned above. This time it brought in different sociological theories of elite formation, from the earlier classifications between positional— and reputation elites to the more contemporary ones, based on the notion of "elite connection" with the public, thus giving the latter more agency. While it transcends the structural rigidity of the prior attempt and explicitly notes that the history of agrarianism cannot be restricted to a history of parties or ideas (here the organizational component of agrarian elites!), it still cannot escape the structuralist frame in which it was conceived.

The fate of agrarianism was very context specific. Hence, the three polarities in Yugoslavia, Bulgaria, and Czechoslovakia are so significant. Internal pressures and contexts determined the expression. As stated above, the most common reason given for failure was weak ideology and weak organization, on the national and international plan. The problem with this lumping together, even granted certain problems and weaknesses, is that most of the time these movements were not given the time and space to mature. The golden age of the European peasantry happened during the most turbulent decades in these countries' modern history. Between repression and the Great Depression, the agrarian parties were prevented from accomplishing the transformations they sought. Even in the exceptional case of Czechoslovakia, there was no insulation from the destruction of the state from without. Johan Eellend quotes from Mary Samal's dissertation on the topic of failure, "All the Peasant parties of eastern Europe were cheated of their legitimate claim to power by the alliance of the crown, army, Socialist parties, and the bulk of the urban population."26

Having said this, three universal initiatives underwrite the agrarian project: (1) parliamentarism, (2) land reform, and (3) the cooperative movement. To see agrarianism in praxis and in relation to these components, the examination of the reform activity of the independent BANU rule in Bulgaria is instructive. All agrarian movements had in common the desire to elevate the peasant and make him the maker of his own history—in a word, to imbue him with agency. While the literature varies in its evaluation of the modernization aspirations of the agrarian movements, my argument here is that there is little doubt about their modernizing potential. Still, it is one thing to gauge this from ideological platforms and another from praxis. Luckily, we have in the short-lived rule of Stamboliiski's BANU (1919–23) the perfect example of ideas put into practice.

Already before coming to power, the BANU had an elaborate ideology and a reform program, whose most important statement was Stamboliiski's book *Politicheski partii ili suslovni organizatsii* ("Political Parties or Estate Organizations").[27] The centerpiece of the agrarian doctrine was the estate theory and the idea of people's rule (*narodovlastie*), as well as the philosophy of "labor democracy." According to the estate theory, society consisted not of classes but of estates grouping people of similar professions, often with diverse interests.[28] At the time, the following estates were thought to comprise the social body: the agrarian estate, the estate of the craftsmen, the estate of hired workers, the industrial estate, the merchant estate, and the bureaucratic estate.[29] The peasant estate was the one that was crucial to the well-being of society with its labor but was held in abysmal conditions because of the exploitation by money-lenders, merchant, lawyers, doctors, and, in general, by the city. Political parties had played a positive role after the passing of monarchical regimes, but they were seen as becoming enfeebled by partisan struggles and also as becoming redundant.[30]

The BANU as the estate organization of the peasants was waging a legal struggle in their interests. The notion of people's rule (*narodovlastie*) was left for the future, except that the form of government would be republican, the ultimate goal being to empower politically the majority of the population (which, implicitly, was the peasantry).[31] The economic basis of people's rule was "labor democracy." State power was the means to implement deep-reaching reforms for the benefit of the small and middling peasants, based on the idea of small property that was managed by individual or family labor, without the exploitation of hired labor. This would bring about "labor democracy" and social justice, achieved through a basic transformation of capitalism, but short of the abolition of all private property. Quite to the contrary, small private property was to be affirmed, but large properties were to be nationalized, and capitalism regulated. Finally, all of this would be accompanied by a broad social program making education, medical care, and the judicial system affordable for the broad masses of the population. This was the essence of the vision of a "third way" between capitalism and socialism.

The implementation of this vision began immediately after the BANU came to power. Throughout its brief rule, it passed more than one hundred laws, not counting a number of purely administrative measures.[32] Bell summarizes the BANU's activities thus: "Labor property, Compulsory Labor Service, cooperation, new forms of education and administration—these were the tools the Agrarians employed in their struggle to transform Bulgaria."[33]

Labor property (*trudova sobstvenost*) was the cornerstone of the land reform. The principle had been elaborated by Raiko Daskalov while in prison in 1916–17. He had been contemplating the decline of the Roman Empire and reached the conclusion that this came about because of the excessive concentration of property in the hands of the rich. To promote an alternative outcome, Daskalov proposed the principle: "The land belongs to those who till it."[34] Accordingly, no one should possess more land than he and his family could cultivate. But everyone should possess enough land to secure self-support.

While Bulgaria, without a landed aristocracy, was the Balkan country with the most equitable land distribution, the growth of the population and the influx of 450,000 refugees after the wars necessitated the introduction of land redistribution. At first, in 1920, a state land fund was established for the distribution of land to landless peasants or ones whose holdings were too small to support them. Maximum amounts were established at 30 hectares of arable land per household, and for absentee landlords this was 4 hectares. Compensations for confiscated lands were paid on a sliding scale by the government. Crown and monastic lands were also affected, and a "Directorate of Labor Property in Land" was set up.[35]

When the BANU was overthrown, the files of the Directorate "contained appeals from 28,325 landless peasants, 74,420 dwarfholders, and 7,500 rural laborers. Of those, the Directorate had dealt with the requests of 17,127 landless peasants, 54,471 dwarfholders, and 4,407 rural laborers. In addition, 18,000 peasant families had been settled on state lands through direct administrative processes."[36] The most remarkable thing about the land reform is that it was kept in place and further put through even after the agrarians were toppled, except that peasants had to pay more for the land they had acquired.[37]

While the land reform was the most radical, the Law for Compulsory Labor Service, adopted on June 14, 1920, was destined to become the most famous. In its initial form, it provided for the conscription of every male over the age of twenty for a year of labor service and of every female over the age of sixteen for six months of service. No exemptions or permissions for substitutions were made. It also foresaw temporary conscription of three to twenty-eight days for men between twenty and fifty years of age. A coordinated assault on the law, by the suspicions of the Inter-Allied Control Commission on Disarmament that this was a way to circumvent the disarmament provisions of the Neuilly Treaty, as well as by the appeals of the Bulgarian bourgeoisie, appalled that their children would be conscripted, and by accusations of the law being socialist, led to a major revision of the law. The "labor army" was reduced, and the purchase of exemptions allowed. When the so-called *trudovaks* ("laborers") were called up in November 1921, of the forty thousand summoned, only thirty thousand actually served. They were employed mostly in road construction, the construction of railways and canals, the building of a communications network, in farming state-owned lands and exploiting state forests, and the maintenance of public buildings and grounds. At the same time, the idea was to instill these young people with discipline, as well as introduce them to skills and basic health and hygiene education.[38]

The Law for Compulsory Labor Service was a purely Bulgarian idea, and its practice endured.[39] In the end, 774,000 people passed through the Compulsory Labor Service. They built 196 kilometers of new highways, repaired another 108 kilometers, built 435 kilometers of village roads and 329 streets, constructed 108 new bridges, and repaired 254 old ones. They repaired wells and fountains, created sewage systems, built cooperative homes, community centers, and schools.[40] As Stamboliiski summarized it, "This law will be redeeming for Bulgaria; it is the foundation of our reforming activity. . . . The ideas invested in this law will be embraced by all progressive people from different countries."[41] Indeed, in the 1930s, President Franklin Roosevelt's administration studied the Bulgarian Labor Service Law as a model for the Civilian Conservation Corps.[42]

The cooperative movement was not a novelty to Bulgarian social life.[43] However, the BANU enhanced the consumer, producer, and credit cooperatives to an unprecedented degree, making them a pillar of its program and a state priority, with the goal to unite all Bulgarian peasants into a national cooperative network.[44] Only through cooperatives could the small and medium-sized peasants take advantage of the technologies and knowledge of the market, available to the large capitalist farmers. Several experimental producer cooperatives were set up in villages in North Bulgaria for the cooperative cultivation of the land.[45]

The most important initiative in this respect was the creation of the Grain Consortium, which was intended to raise and stabilize the price of grain. It was given monopoly of grain exports and offered set prices to the peasants, higher than the ones offered by private traders. It succeeded in ending the rampant speculation, although the private grain dealers managed to persuade the Reparations Commission to disband the Consortium on the pretext that it set grain prices at artificially high levels.[46]

Yet, the cooperative venture persisted in other branches, such as fishing, forestry, livestock, rose oil, or wine growing. Most importantly, it made huge inroads in the flourishing tobacco industry. The tobacco cooperative *Asenova Krepost* in Stanimaka (Asenovgrad) became "a self-proclaimed Jerusalem for the cooperative world, even going on the road to display its concept at international venues such as a trade fair in Rio de Janeiro and an international cooperative exhibition in Ghent, Belgium, in 1923."[47] Again, even in the circumstances of diminishing state support after the ouster of the BANU, the Right-wing governments that followed saw in the cooperatives a stabilizing force regulating urban-rural relations and a bulwark against communist influence, that was especially strong among tobacco workers.[48]

Designed as an experiment between the local cooperatives and the state, at the end of 1920, the BANU launched a major project to develop Bulgaria's water resources. The legislation was based on the Prussian law but organized on a cooperative principle. Local cooperatives, called water syndicates, were formed for irrigation, to produce electrical energy, for drainage, the strengthening of river banks, and so on. Major rivers were to be used for irrigation and for the building of dams. Thermal plants were to be built next to the mines. The final goal was to create a unified electricity net for the whole country that would not only promote agriculture but endorse also the chemical and metallurgical industry.[49]

> In twenty years Bulgaria will become a model for an agrarian country, whose towns and villages will get rid of their crooked and muddy streets, and of the people's bloodsuckers. They will be supplied with clean and healthy drinking water, with wooded parks, with modern chemical fertilizers, with telegraphs and telephones and electricity. They will have highly developed cooperatives, a broad railway network with wheat and tobacco depots at each station. Each village will have its House of Agrarian Democracy, where lectures will be held and films will be shown, and where the peasants will be able to hear recordings of the best speeches of the best orators.[50]

This bold vision gives the lie to the accusations that the BANU, like agrarian parties in general, in its opposition to the excesses of capitalism, was against modernization.[51]

The cooperative initiative was extended also to the urban sector. New apartment blocks were built, financed by cooperatives but with private ownership of the apartments. The BANU government set norms, allotting each family two rooms and a kitchen, with provisions for larger households. This was understandable, given the housing shortage and the influx of refugees, coupled with the minimal housing construction during the war.[52] Yet, this was one of the most unpopular reforms of the BANU government and was especially resented by the urban population, which felt that it had been overtaxed in favor of the countryside.

Indeed, in its fiscal legislation, the BANU attempted to shift the burden away from labor property to other spheres. Traditionally, the bulk of Bulgaria's revenue came from the tax on land. While the agrarian program anticipated a unified progressive tax, the government did not dare introduce it. Instead, the BANU introduced a sliding progressive income tax reaching a maximum of 35 percent. The government gave preferential treatment to cooperative enterprises, shifting the tax burden toward joint-stock companies.[53] The fiscal legislation of the BANU incurred huge protests from the financial-industrial circles, as well as from its own Right wing, and, as a whole, remained unpopular and weakly implemented.

Throughout the whole period of its rule, the BANU had tense relations with the communists who opposed most of their legislative measures. Still, it would be unfair to say that the BANU was unconcerned with the plight of the working class. Not only did the agrarians introduce the eight-hour working day, but they believed it unjust that workers had no share in the factories. As Bell reports, "Legend has it that Stamboliiski was at his villa preparing legislation to this end when his government was overthrown."[54]

The reforms in the educational sphere were directed primarily to the goal of creating more and better prepared professional and practical cadres. What was needed, in the view of the agrarians, were more engineers, artisans, and better educated farmers. Stoian Omarchevski, who became Minister of Education in May 1920, had studied the educational systems of Great Britain, Germany, and the United States, and was particularly inspired by the ideas of John Dewey.

Education was made more accessible, especially in the countryside. Until the BANU came to power, Bulgaria had only four years of free compulsory elementary schooling. The agrarians added another three years in the middle school. Huge funds were disbursed for the building of schools, and several hundred new elementary and middle schools were opened between 1920 and 1923.[55] The main effort was the change of the curricula of the middle (*progimnazii*) and high schools (*gimnazii*), and their infusion with more practical subjects, such as agriculture and applied science. The traditional five-year high-school program, heavy with literature, history, and languages, was divided in two parts. The first three years (*realka*) stressed work-related education, adapted to local conditions. Thus, in rural areas, they studied agriculture; in towns, handicrafts and industrial arts; in wooded regions, forestry; and so on. The remaining two years followed with traditional subjects in view of preparing students for the university, or else, students could enter professional schools and institutes.[56]

In the realm of higher education, faculties of medicine, veterinary medicine, and agronomy were added. One of the major reforms, as was detailed in the previous chapter, was the orthographic one, which was repealed at the end of the agrarian rule, to be adopted again by the communists after the Second World War. All in all, the financial support of the University, of the Academy of Sciences, and of libraries was increased. However, the agrarians undertook a number of unpopular measures that further alienated the opposition both from the Left and from the Right.

While some of the reforms have been deemed unsuccessful, such as the health reform, they did leave a deep imprint on Bulgaria's social policies "by introducing into Bulgarian politics the notion of preventive medicine and positive eugenics, while also touching upon the sensitive problems of child mortality and maternal healthcare."[57] According to the Minister of Interior and People's Health, Raiko Daskalov, who introduced the new law:

> It is only the radical reorganization of our healthcare legislation which would yield the desired results expected by the state's investments—a healthy and strong nation, capable of meeting all the challenges of civilization and preserving its independence amongst the culturally elevated counties of the world. Any merely bureaucratic approach toward healthcare has to be dropped on the spot and instead, the slogan of public health should be raised: "Healthcare comes as a priority; the people's health comes above all!"[58]

The Bill for People's Health was placed before parliament for consideration in February 1923, was sent back after the first hearing, and was not heard of again. Remarkably, however, the agrarians were the only government in Bulgarian history to have prioritized the funding of the Directorate for People's Health above that of the police.[59]

To complete the task of this chapter which engages with the problematic of agrarianism in the context of its aspirations and limitations, I would like to present an annotated summary of the first three years of the publication, starting from 1923, of the Green International, the *Bulletin Mezinárodního Agrárního Bureau*. In Chapter 2, I discussed the reorganization of the MAB as a result of Mečíř's appointment to head the organization in 1926. Subsequently the MAB abandoned its Slav orientation and became a pan-European organization. At the time, in order to mark that transition better, Mečíř suggested that the MAB had almost ceased functioning and that this dysfunctional state had penetrated into the *Bulletin*, so that even the simple task of publishing it regularly failed. This picture is a caricature, as the following survey will show.

The merit of this exercise in presenting the work in the *Bulletin* lies not only in the fact that it has never been done before. Presenting its work in this way will show the varied initiatives and coverage it engaged in and will correct the various holes, misrepresentations, and errors in the scant literature. For the very few works that have touched on the work of the Green International, the organization is either of tangential interest,[60] or in the single monograph that covers it, it is so skewed through ideological polemics that it becomes unrecognizable.[61] Daniel Miller's meticulous work on the Republican Party unfortunately does not connect to the MAB. He is incorrect in stating that the project was put on hold until 1927 or that its first conference occurred only in 1929.[62]

The first issue of the *Bulletin of the MAB* (1923) opened with a statement printed simultaneously in Czech and French. The Czech title translates into "The Idea of World Agrarianism [*Idea světového agrarismu*]," while in the common language of diplomacy, but also the radical language of the French Revolution, the title expands a step further to, "The Idea of Universal Agrarianism [*Idée de l'agrarisme universel*]."[63] It is a credo that extols the virtues of the land and its tillers. While there are numerous references to the land as the breeding ground for all life and then by extension for civilization itself, this aspect is very quickly hitched to a programmatic vision of the peasant as a political subject. "Thus the man living on the land is and must be the creative element in the state."[64] Without a proper engagement with the peasant, "the world situation will only improve when Europe starts taking care again of the source of its existence, [that is] takes care of agriculture."[65] The other element to this equation is the peasantry's peaceful contribution to the development of the world, which serves as a counterpoint to the destruction of the First World War. The rather preliminary programmatic document concludes nonetheless with a manifesto:

> It is incumbent for the agriculturalists of the world to unite,—while understanding their importance and the value of their common and lawful destiny and of their collaboration,—for the good of humanity, for the defense of the social order, for the support of the state on the road of peace, and for the security of agriculture, that is to say, to fulfill the agricultural idea and to give humanity, to states and to nations, by means of food and by the nature of their [agriculturalists'] existence, the firm foundation of life, well-being that is physical and moral.[66]

The major articles in the rest of the MAB publication for 1923 included informative pieces on the effects of the Russian Revolution on Russian Agriculture (émigré

Emelianoff), on the Yugoslav agrarian movement (Stajič), on the BANU (Obbov in emigration in Prague), and on Poland (Bader). The issue included a long article on the other important component of agrarianism, the cooperative movement, titled "The Agrarian Cooperative Movement in the Slavic World." There was a report on the first session of the consultative agricultural commission that was part of the International Institute of Labor at the League of Nations by Ferdinand Klindera, the president of the *Centrokooperativ*, the successor to the Central Union of Economic Cooperatives.

Only two pages were devoted to the coup d'état in Bulgaria, and that report was sandwiched between the cooperative article and the Economic and Industrial Exposition in Gothenburg. Cooperation was treated again in its Romanian context. There were even informative articles from a broader context: The American Farm Bureau Federation, for example. After a rubric on the agrarian press there was a closing section that contained a list of noteworthy events per country: Bulgaria, Czechoslovakia, Yugoslavia, Germany, Poland, Russia, the United States, and Sweden.

Nothing in that first year of publication hints that the MAB could be anything more than a platform for the exchange of information about what is going on with the member countries.

This impression is reinforced by the production from 1924. The MAB could only get out one issue (January–March). Švehla contributed an article that mirrored the one in 1923 on the soil and peace. Among the general reports was a piece that described Turkey as an agricultural country and another on the Swiss Peasant Union. A longer article on agrarianism by the Russian Social Revolutionary, Pitirim Sorokin, was in essence a set of sociological ruminations on agrarianism that concluded with, "Agrarians of the whole world unite." Finally, there was a longer piece on "parliamentarism and democracy."

In the rubric "news from around the world," the Bulgarian section only covered the grain harvest and the 1923 budget, the Czechoslovak covered the "agricultural week" that included practical courses for agrarians (March 2–8), the formation of a Student Agrarian Club of South Slavs in Prague, and the life of the Russian agrarian émigrés. For Yugoslavia, the topics were the State Agricultural Bank, the agrarian question in Bosnia, Agricultural Education, and Health Cooperatives in the countryside.

However, not only did 1925 produce three issues, but the first issue reported on and filled in the activity in 1924. The major components of agrarianism, parliamentarism, land reform, cooperatives started being fleshed out after having been only touched on in 1923 and 1924. Issue One started with a detailed treatment of the Land Reform in Czechoslovakia by Karel Viškovský, the president of the State Land Office. It treated the importance of land redistribution, the content of the reform, the results, and also colonization. His conclusions justified the need for the reform, its positive effects, and also the compensation of expropriated landowners based on prices before the First World War. He described the reform "not as an act of social or political vengeance, [but] it is based exclusively on social and economic considerations."[67] Land reform was also covered in Yugoslavia and Rumania, but not in Bulgaria. There was an article on Polish agrarian parties and another on the coup in Bulgaria and the fact that it could not destroy the BANU.

This issue was most important for its report on the 1924 conference of the MAB during the agricultural exposition in Prague. It was attended by the *Savez Zemljoradnika*, the Polish People's Party "Piast" and the Polish People's Party "Wyzwolenie," the BANU, the Republican Party, and the Russian agricultural émigrés, headed by Ivan Emelianoff, who before the revolution had been the chief of the Economics Bureau of the Provincial Zemstvo of Kharkiv and then studied economics in Prague from 1921 to 1924.

The MAB describes the conferences as being, "in one part, a review of the work that the Bureau had accomplished . . . such as mutual convergence [re. the agrarian parties (*ve vzájemném sbližování*)] and in the comparison of the programs and methods of work of the agrarian organizations, as well as, in the other part, in the gathering and study of the materials and documents of the agrarian movement, its manifestation among all nations, and the tracing of new paths that lead to the fulfillment of the demands of agriculturalists."[68]

The sessions of the conference began with a dedication by Švehla to the oppressed agrarian brothers in Bulgaria and Russia, which produced a unanimous statement of opprobrium by the secretariat. Obbov spoke against the Tsankov terror, V. Lazić (*Savez Zemljoradnika*) on the principles of agrarian organizational activity and the necessity of an agrarian organization for all the Slav nations, the representatives of the Polish parties, M. Downarowicz (*Wyzwolenie*) and L. Skulski, and A. Erdmann (*Piast*), spoke about the necessity of agrarians to become a force for peace in the foreign policy of Poland and Czechoslovakia, Emelianoff on the problems of the Russian agrarian emigration, Švehla on the agrarian idea, Klindera on the possibility of international relationships for the cooperative movement, and F. Stanek on the necessity of publications on life in the agrarian countries and their structures.

The conference was summed up thus: "All the reports of the delegates to the conference provide **proof of good unanimous will of all Slav Agrarians**

**to work in order to achieve, by the deepening of estate awareness, national education, and constructive politics, their share of power in the domestic and international politics of their states, where the peasant democracy has a duty to introduce the principle of its natural pacifism and consolidation** [emphasis in the original]."[69] The Czech version uses "deepening of estate awareness" (*prohloubením stavovského uvědomění*) whereas the French translation changes this to "deepening of class consciousness" (*l'approfondissement de la conscience de classe*). That is telling because the Czech version reflected the ideological position of Stamboliiski's estate theory that he had developed in his 1909 book *Politicheski partii ili suslovni organizatsii* even while his contribution to the formation of the Green International was being downplayed if not actively erased.[70]

The conference concluded with an appeal:

> All the people involved in agriculture the world over: . . . In this century of civilization and humanist ideas, after an era in which false social and political doctrines only brought humanity demoralization, famine, and death, humanity needs to draw new moral and spiritual strength from the hard work and spirit of the agriculturalist, this regenerator of society. *This precisely is the meaning and problem of agrarianism—to convoke all the agriculturalists of the world to common work* [italics in original], awakening their awareness of their importance and force, requiring for them a fair share in domestic and international power.[71]

This formulation of agrarianism was synonymous with its function in agency formation and the elaboration of subjectivity that this monograph has been advancing.

The *Bulletin* then reported on the constitutive assembly of the Union of Slav Agrarian Youth (*Svaz slovanské agrární mládeže*) in Ljubljana in September 1924. It included organizations from Bulgaria, Yugoslavia, Czechoslovakia, and Poland. The preparatory commission began work in 1923, first in Prague, then Warsaw, and finished in Ljubljana. There was also information about Lauer's Swiss project, and the criticism about the organization is very muted. The Russian émigré component still found coverage in the *Bulletin*, in this issue concerning the creation of a Cossack organization. The information section's report on Bulgaria described the difficult position of the Permanent Presence of the BANU and reported on the mission of the Representation of the BANU Abroad in Prague as the true voice in front of European opinion on the situation in Bulgaria. Further, it informed that the Representation Abroad had twice rebuffed efforts of the *Krestintern* to enter into relations with it. For Czechoslovakia, the report focused on the creation of the Agrarian Credit Cooperative (*Zemědělské sdružení pro opatžování úvěru*) as a continuation of the logic of the land reform. Also, on December 28, 1924, the Agricultural Academy in Czechoslovakia was founded, which was the second academy of its kind in Europe after the one in Paris. The report on Yugoslavia included an article that criticized Radić and Pašić from the perspective of the *Savez Zemljoradnika*. In the elections for parliament, they had dropped from 160,000 votes and eleven mandates in 1923 to 126,000 votes and five seats, and the report analyzed the causes of this disappointing result. In sum, even though the *Bulletin* had not appeared regularly in 1924, the subsequent impression that the organization was not functioning is completely incorrect.

The second issue of 1925 began with the important article of Švehla "The Land and the State," which was in essence a rehashing of his ideas in the Party Program of the Republican Party. It was followed by a long article entitled "The Agrarian Movement as a Factor for Yugoslav Unity." It was a critique of Radić and presented the *Savez Zemljoradnika* as an agent of national unity. Articles from Russian émigrés were still included due to their pan-Slav orientation.

A critique by Mečíř prefaced the coverage of the XII International Congress of Agriculture in Warsaw. The professional orientation of the Warsaw congress was counterposed to Mečíř's and Fiedler's statement at the MAB Congress: "By agrarian politics we wish to be understood not only the politics made at the Ministry of Agriculture, but also the general politics and all its branches, for it is obvious that there isn't one that does not touch the vital interests of the peasant."[72] They took a stand against neutral, professional/economic associations and demanded real politics that reflected the status of agrarianism and the central role of the peasant in society. They continued: "In our opinion we must say to agrarians: Your special organizations are a great thing, but they lack a summit from which you can fly the flag under which the fighters for your claims can gather." It is a sign of its intellectual integrity that the issue printed the work of the congress and the details of its five sections.

The report of the Second Conference of the MAB that took place again in May during the agricultural week showed the activity of the organization. The attendees were the same as in the previous year, but for the absence of *Wyzwolenie*. The conference was concerned with organizational, economic, cultural, and propaganda questions. In addition, reciprocal excursions were organized for agrarians so that they could visit different countries, as well as for students of agrarian theory and practice. The conference elaborated the stand of the MAB concerning the Congress of Agriculture in Warsaw as well as Lauer's agricultural organization. The congress also approved resolutions to collect information on agricultural statistics, the propagation of the idea that aid should be given to the Bulgarian agrarians, to begin preparation for the next congress, and to prepare a Slav agricultural exhibition.

Another article reported on the law from June 25, 1925, to extend insurance to the self-employed for injury as well as to provide retirement, and its implications for agrarians was discussed. A long article by Obbov described the Bulgarian Agricultural Bank and stated that its origins were almost cooperativist (he was referring to its origins from the agricultural cooperative banks during the Tanzimat reforms of Midhat pasha). Obbov treated the Stamboliiski reforms in relation to the bank. The article was intended to serve as a proof of the utility of cooperation and made the case through the example of the Bulgarian tobacco industry. The issue also welcomed the easing of the agricultural policies in the Soviet Union starting in 1925 and then concluded with an analysis of the land reform in Yugoslavia.

The third and last issue for that year started with a letter to the MAB by Gabriel Fleurant "Agricola" from France who would a few years later found the *Parti agraire et paysan français*, which in 1934 joined with the *Comités de défense paysanne* of Henry Dorgères to form the *Front Paysan*.[73] The similarity in positions between Fleurant and those advanced by the Republican Party were warmly welcomed. The issue also reprinted Švehla's speech at the Republican Congress in 1925, "Agriculture, War, and Peace." The arguments were that regeneration after total destruction comes from agrarian democracy. The goal was to avoid war: "To achieve this goal, the agrarian/peasant has to achieve power in all countries and to create a state of affairs which will bring to the world peace, and through peace—work, bread, liberty."[74]

The issue included an article on the Land Code of the Soviet Union from October 30, 1922, and its supplement from March 29, 1923, which was very scientific and measured. Also, it contained a theoretical piece on international politics and the international agrarian organization by Hodža. Hodža argued that there was nothing besides agrarianism that could unite the Slavic peoples with the chief goal being the consolidation of peace. "And I ask you, which principle, which economic or social program could penetrate into all the classes of the Slav peoples so profoundly so as to unite them, to form a bridge between not only Slav people, but also non-Slavs, so as they have the same interests."[75] Neither nationalism, nor clericalism, nor socialism could accomplish this work that Hodža terms the consolidation of Europe! In one of the first open broadsides against Bolshevism by the MAB, Hodža wrote: "For the moment, we must consider the Bolsheviks as a permanent menace directed both against agrarianism and pacifism which should become the principle of international politics."[76] He concluded that despite a softening in the USSR, exemplified by Bukharin, the sole policy of the agrarians was to have a distinctive front and no cooperation.

Another impressive contribution to the issue was the report that Otokar Frankenberger from the Ministry of Agriculture presented to the Czechoslovak academy of agriculture, "New Agricultural Policy." It was a criticism of the spirit of consumption and the liberal economic orthodoxy that the best return on investment (i.e., private interest) translated into the best productivity (the public good). His counterpoint was that the conception of the public good could not be abandoned to the invisible hand, but that as an ideal it depended on moral and cultivated individuals, a democratic people whose members were equal in their rights, and that prevented the supremacy of a class, whichever that is. Agricultural economic policy, in order to achieve this, had to be interventionist, in other words, political. And that demanded an etatist intervention to stop agricultural exploitation through price and tariff policy.

Georgi Vulkov then contributed on the condition of the BANU, the achievements of Stamboliiski, and the repressions. Despite the adopted position of criticism for the project of Lauer, the *Bulletin* also reported on the International Conference of Agricultural Associations in Berne that took place from September 23 to 24, 1925. It even printed the articles of the association. Before concluding with the news section from around the world, the issue also contained thorough coverage of the Congress of the Republican Party at the time of its twenty-fifth anniversary.

Even from this brief recital of the contents of the *Bulletin*, one can clearly see the amount of activity undertaken in the first three years of the Green International's existence. In the subsequent years, the *Bulletin* continued to inform with the same high standard on the activities of the MAB and on the state of agrarianism. The MAB was never a revolutionary institution, but that does not mean that its informative thrust and its politics of soft power justify its dismissal. The *Krestintern* had a tight organizational structure, the support of the *Comintern* and the Soviet state behind it, as well as the partnership of the national communist parties, and still it could not break into the village. If the Green International did not adopt more aggressive tactics, it was because of the constant belief that agrarianism was viable and could offer a solution not only to the peasants but also to the world.

This chapter established the tension between the aspirations of agrarian political reform and the constraints that hindered the implementation of these aspirations. The agrarian moment took place during some of the most brutal decades of European history. War, reconstruction, the Great Depression, and another war, provided a baseline to a tune of repression, challenges to parliamentarism, coups d'état, and dictatorships. The resilience of agrarianism through these challenges, I argue, is an eloquent testament to its strength.

The chapter looked at several moments of astonishing strength that normally have been treated as moments of weakness in the literature. The reason for this erroneous assessment, I argue, was due to the excessive focus on ideology in the study of agrarianism. As a result, my critique was differentiated from the classificatory approach to agrarianism in classical studies of populism as well as from the recent effort to synthesize agrarianism in East-Central Europe that is hindered by a structuralist frame. While I argue that the fate of agrarianism was context-specific and that it produced the three faces of agrarianism in the national contexts that are the site of this book, I still propose three universal initiatives that underwrite the agrarian project: parliamentarism, land reform, and the cooperative movement. As an illustration, this chapter thus illuminated these initiatives in the radical agrarian phase of Bulgarian history. The next chapter returns to the mode of a case study in order to mine one example of the repression exerted over the agrarian parties, that of political delegitimation through charges of treason and corruption. The more brutal tactics of political murder, White Terror, and police repression are better known, so this example of "soft" power is useful to flesh out the broad array of tools that the opponents of agrarianism employed to constrain it.

# 5

# Delegitimizing the Agrarian Alternative: The Diptych of Stamboliiski's Corruption and Radić's Treason

The inhospitable climate of the interwar period tempered the ambitions of political agrarianism and accreted a structure of limitations that prevented the full expression of the movement's aspirations. If agrarianism's success is to be measured by the achievement of its maximalist program, then the failure to achieve this was clearly overdetermined. On a regional and pan-European level, the steady turn toward authoritarianism eroded the parliamentary platform on which agrarianism depended. Similarly, the Great Depression devastated agricultural prices and upended the calculus that modernization could be financed through equitable agricultural development, while at the same time favoring economic autarky. Factionalism in the agrarian parties over how to address those challenges weakened the movement, even before political oppression and White Terror further splintered it. Yet, this generalized picture is better left in the background if one seeks to explore the specific limitations faced by each national agrarian movement. The chronology and precise factors affecting each agrarian party differ sufficiently to necessitate individual examination. Fortunately, the national literatures on agrarianism are sufficiently developed with respect to these "hard" limitations. Not as developed, however, are the "softer" initiatives to contain and delegitimize the agrarian alternative. The introduction of cultural criticism and analysis to this problem is of enormous methodological utility as it not only allows the introduction of new sources but also permits the recontextualization of events and developments. That, in turn, opens up the possibility of relation to historiographic and theoretical debates outside of the Balkan and East-Central European regions—for example, in the analysis of the underlying ethics of competing political systems or in the exploration of the mechanisms of induced cooperation.

This chapter examines two such interventions to delegitimize the agrarian movements of Bulgaria and Croatia. By manufacturing a charge of venal corruption against Stamboliiski, Tsankov's regime attempted to fundamentally discredit and erase the systemic alternative offered by the BANU's reforms up to 1923. Profiting from Radić's trip to Moscow on June 2, 1924, where he joined the IPC, the *Krestintern*, the Belgrade government proclaimed the *Obznana* against the CRPP through the law for the protection of the state on January 1, 1925. The charge of treason against Radić encompassed the CRPP as well and virtually outlawed it. As Mark Biondich writes, "The party was to be suppressed, its leadership jailed, its newspaper banned, and its archives seized—in other words, it was to be placed outside the law."[1] Thus, this chapter explores the articulation of these delegitimation campaigns, analyzes the responses to them, and comments on their broader theoretical and historical significance.

The daily newspaper of the Democratic Party, *Priaporets*, began its first issue since the coup d'état against the agrarian regime of Alexander Stamboliiski on June 9, 1923, with the following lines: "A regime of bloody and disgusting tyranny, the memory of which will burn the clear forehead of the Bulgarian with shame, collapsed under the pressure of its crimes and unheard of debauchery."[2] Only three issues earlier, this same newspaper published a front-page article entitled "In Cultured Countries and in Ours." The article related how an Italian deputy who had spoken against the repressions of the fascist government had been severely beaten upon exiting parliament and how the whole political establishment had denounced this episode. To mark the difference in cultures between civilized Italy and the tyranny of the Bulgarian peasants, the author posed the rhetorical question, "Imagine if instead . . . of Mussolini it was somebody [such as] Stambol[iiski] from *Zemledelsko Zname*, what kind of defense would this poor deputy have received inside or outside of Parliament?—Yes, we are Europeans, says Bai Ganio [a literary figure which personifies Bulgarian provincialism], but we are always not quite there yet."[3] With little variation, this was the general tone of coverage in the news organs of the traditional political parties that had been in opposition to the BANU regime. While these were easily lumped together by communist historiography under the rubric of "the bourgeois press," these newspapers did not just represent the parties that constituted the Democratic Alliance that seized power on June 9, 1923, but also the socialist press. From the Center-Left to the Far Right, therefore, the coup was euphorically welcomed. The language announcing the fall of Stamboliiski's regime was striking in its avoidance of referring to the actors and actions that toppled the regime. Instead, the extensive use of the passive voice makes it seem as if the "*druzhbashi*," the pejorative appellation used to refer to the united agriculturalists, caused their own fall.[4] In order to effectuate this rhetorical sleight of hand in which the coterie of officers and Macedonian autonomists that sadistically hacked to pieces the prime minister of a lawfully elected government disappeared and was replaced by a group of "corrupt" despots that brought about its own downfall, the press committed its resources to describing the purported moral degeneracy in all spheres of the agrarian regime. This can be perfectly seen in the inversion of the famous chastisement of the sins of the city by Stamboliiski by the Socialist Party's newspaper, *Narod*, which included an exposé entitled "From the *druzhbashki* Sodom and Gomorrah."[5]

## A micro-history of the corruption trial

The keystone in this concerted discrediting of the BANU regime was the personal corruption of Alexander Stamboliiski. By June 13, the site of the "Sodom and Gomorrah" had narrowed specifically to the house of Stamboliiski in Sofia.[6] The initial reports were sensationalist lists of "compromising" materials found in Stamboliiski's residence that supposedly showed him as one of the richest men in Bulgaria—one that lived like a king. Boxes of condoms, pornography, and heaps of silk female lingerie attested to his lasciviousness, while a myriad of pomades, creams, and perfumes were the proof of his vanity. Most important, however, were sums that included 1.25 million Swiss francs, 9 million French francs, and 1 million Bulgarian leva. Based on the current exchange rates, these sums were valued at twenty-three million leva.[7]

A tendentious announcement by the Minister of Finance was published in every pro-coup paper between June 13 and June 15, 1923. Although the titles differed slightly from case to case, the fact that the text was identical suggests that it was printed verbatim from an official source provided for the purpose. While the statement began with the announcement that the Cabinet of Stamboliiski had budgeted yearly six million Swiss francs for the "protection of the cause of the fatherland," but which, instead of going to foreigners, was used by the people close to the former ministers to "reap a richer yield," it quickly moved to the personal misdeeds of Stamboliiski.[8] The account stated that Stamboliiski used a top secret decree No. 737 of the Council of Ministers to receive 4 million Swiss francs, the equivalent of 120 million leva, on March 15, 1923.[9] Suspicious assessments abounded: "When he [Stamboliiski] ordered that this operation be known only to three clerks who were to execute it, he told to some of them personally, 'I will treat any disclosure equivalent to treason,'" but the reported "facts" were as follows: four checks were prepared for the Banque Fédérale Genève, Société de Banque Suisse, Credit Suisse, Union de Banque Suisse; Stamboliiski's son-in-law, Ivan Boiadzhiev received them on March 18, 1923, and left for Switzerland; he returned on April 16, 1923, with Swiss banknotes; he was escorted by security from the railway station to Stamboliiski's homestead, from where the money disappeared.[10] The final summation of this "unprecedented theft" was that for a country the size of Bulgaria, it even "exceeds the Panama Affair."[11]

The delegitimation of the agrarian regime particularly through the lens of Stamboliiski's malfeasance continued through the summer of 1923, although the only additions toward the end of June consisted of an inventory of all the banknotes found in his apartment. Amid the different currencies found, the more substantial sums consisted of about 2 million Swiss francs, 200,000 French francs, and 2 million Bulgarian leva.[12] The September Uprising and the atrocities in its suppression marked the end of the preoccupation with Stamboliiski, at least in the press. However, the propaganda machine of the Democratic Alliance, headed by Alexander Tsankov, maintained the offensive until the end of 1923 through the publication of a remarkable series of brochures. Entitled *Druzhbashkiia Rezhim—Dokumenti*, the twelve volumes in the series were the work of an anonymous, but grandiosely titled, Commission for the Examination of the Archives of the Former Ministers.[13] With the exception of the first two, which are devoted specifically to Alexander Stamboliiski, each of the brochures treated a topic designed to defame the deposed BANU regime. Some reflected the evaluations in the foreign press about the "ministerial change of the 9th June" or about the BANU regime in general; it goes without saying that only negative ones were included as proof of the opprobrium of the entire civilized world.[14] Others, such as "Partisan Gangrene in the School," compensated for a scarcity of documents through rhetorical flair.[15] While the authors took pains to document their revelations, this effort rarely followed any conventions for citation. Only the sustained invective held together the jumble of quotations, transcriptions, and facsimiles. In addition, the most damning of these "documents" could not be found anywhere else, which could lead one to conclude that they were either forgeries or discarded by the commission once its work was done. This makes it extremely difficult to evaluate this corpus as a source. One thing is clear, however. As a piece of propaganda, the series was a state-of-the-art production.

Only the first two brochures on Alexander Stamboliiski are of concern here.[16] The sections "debauchery," "baronial life," "scams," "poverty and riches," and "thefts" are recapitulations of the scandals in the press. Together, they amount to half of the volume of the two brochures. The last section, "thefts," however, is the most massive and best documented, for it is on corruption that the final dismissal of Stamboliiski and his political project hinges. Its conclusion was clear: "Despite his loud boasts about 'breaking the ice' and 'widening of the horizons,' despite the numerous 'predictions' for 'imminent Spring' and 'a rich harvest,' Stamboliiski was leading our fatherland toward a catastrophe—political and economic catastrophe,—which would have been more horrible than the catastrophe during 1918."[17] As evidence for the catastrophe, the authors maintain that Stamboliiski withdrew 147,186,460.46 leva from the treasury between February 2, 1920, and March 16, 1923. Out of these, 120 million directly referred to the 4 million Swiss francs; the remainder reflected a long list that was provided by the Ministry of Finance concerning payments made to or at the bequest of Stamboliiski.[18] In this category, for example, one finds the travel expenses of Stamboliiski or various payment orders that most likely refer to normal administrative activity. Without enough data to allow the reader to determine for what these sums were intended, let alone whether they reached their destination, their inclusion was meant only to suggest that the prime minister used the treasury as his personal bank.

The brochure was also significant because it provided a final accounting of all the sums found in the house of Stamboliiski in the village of Slavovitsa, his Sofia apartment, and in his cowshed near the capital. The tally was prompted by injunctions No. 4214 of the Ministry of Finance from June 25, 1923, and No. 16583 of the Central Directorate of the Bulgarian National Bank from June 26, 1923. The resulting act of June 26, 1923, was issued by a committee composed of Bulgarian National Bank officials, the representative of the mayoral office, the director of the Budget Office, and a judicial magistrate from the Sofia regional court.[19] The committee counted 1,941,200 Swiss francs, 187,050 French francs, 15,500 Czech crowns, 620 British pounds, 20 dollars, 19,160 lei, 3,672 German marks, 98,000 Austrian crowns, 909 Hungarian crowns, 340 Polish crowns, and 2,002,067 leva. Other valuables included a gold watch and about 200,000 leva in stocks and in bank accounts.[20]

All the property of Alexander Stamboliiski was seized on the grounds of his corruption as it was elaborated by his detractors above. A general sequester of all the movable and immovable property of all the former ministers from the Cabinet of Stamboliiski was announced by the Ministry of Finance with letter No.18492 on the June 25, 1923.[21] However, the political instrumentalization of the corruption charges against him demand some verification, at the very least. To some degree, this can be done through an analysis of the responses that the agrarians gave in their own press. Additionally, the archival records of the

Bulgarian National Bank can corroborate the sums in question. Records from the People's Courts after September 9, 1944, can contribute testimony from the individuals who were involved in the coup of 1923. The most important documentary base, however, consists of a parliamentary bill from March 22, 1941, for the restitution of Stamboliiski's property to his inheritors and the legal proceedings which followed. The juxtaposition of these sources to the propaganda of 1923 does not exonerate Stamboliiski. Instead, the competing ethics of two systems come into focus.

## The agrarian response

The coup, the arrests that followed, and the conditions of the White Terror meant that the organ of the BANU, *Zemledelsko Zname*, was only able to resume publication in Sofia after February 16, 1925. In the meantime, *Zemledelska Zashtita* under the editorship of Kosta Tomov became the temporary organ of the BANU from its first issue on December 18, 1923. This paper was shut down on June 10, 1924, by governmental order no. 25237 because issues 67 and 68, marking the anniversary of the coup, had been dedicated to all the murdered agrarians.[22] Occupied with responses to much more serious attacks and faced with the challenge of reestablishing the crushed organization of the BANU, the paper only responded once to the corruption charges against Stamboliiski: "The millions found at the home of the late prime minister Al. Stamboliiski are state funds, and not stolen money, as the unscrupulous government people claim. Issued properly with an act from the Ministerial Council No. 1 from March 15, 1923, Protocol No. 26, were 4 million Swiss francs or 120 million Bulgarian leva, of which a part was spent for government purposes and the remainder was found. For the spent sums the deceased kept a record."[23]

Outside of Bulgaria, the Representation of the Bulgarian Agrarian National Union Abroad was under less direct pressure, even though its head, Dr. Raiko Daskalov, was gunned down in the streets of Prague on August 26, 1923. The Prague edition of *Zemledelsko Zname* began publication on September 15, 1923. The paper's chief goal was to rally public opinion against the Democratic Alliance and to end the depredations against the BANU. Indignation over political assassinations and repression took precedence over the smear campaign against Stamboliiski. Nonetheless, the paper printed a direct response to the *Druzhbashkiia Rezhim* brochures in the December 14, 1923, article "Why Was Stamboliiski Killed?"[24] The article adopted the position that the baseness of the accusations in the brochures is a reflection of the baseness of the killers of Stamboliiski: "The brochures are an accurate representation of how the killers of Stamboliiski view the 'crimes' of the agrarians as well as of the scale of shameless slander, which they employ as proof for their positions."[25] Even though the article directly addressed only the first two brochures on Alexander Stamboliiski, curiously, there was not a word about the dominant charge of corruption. Instead, the focus was on Stamboliiski's activities during and shortly after the First World War. In a paradigmatic example of specious logic, *Druzhbashkiia Rezhim* faulted Stamboliiski for continuing to engage in the organizational activity of the BANU while incarcerated as a political prisoner for his opposition to the World War. Not only did he not honorably serve his sentence quietly, but his activities in prison and his support of the Radomir Rebellion were molded in that brochure into a variant of the stab-in-the back thesis! The *Zemledelsko Zname* article gleefully dismantles this travesty and on that basis answered the question with which it began: "Al. Stamboliiski was killed for the cause of Panslavic unity, for the cause of peace, for the cause of progress."[26] The focus on these projects was indicative of the priorities of the agrarians concerning the defense of particular aspects of Stamboliiski's legacy; corruption was as much a non-sequitur as pornography.

Nevertheless, *Zemledelsko Zname* wrote on two occasions in 1924 about the found sums. In a special issue from June 14, 1924, that was devoted to the commemoration of the first year since Stamboliiski's assassination, the paper points a finger at Captain Harlakov, in whose custody Stamboliiski was killed. The paper charged that a bag containing several million leva belonging to the BANU was taken by Harlakov and was used by him to build a house in Sofia.[27] The other banknotes were state funds, the paper maintained, and Tsar Boris supposedly knew about them and ordered that Stamboliiski be not persecuted for this matter. The appeal to the authority of the king in the same sentence, in which the paper referred to him as *Boris Posledni*, Boris the Last, is jarring and cannot be verified.[28] On December 15, 1924, a short, solitary announcement proclaimed: "The Representation Abroad has prepared a full exposition concerning the ministerial decree with which four million Swiss leva [sic] were released to Prime Minister Stamboliiski for special purposes."[29] This exposition was never printed and the curious and completely decontextualized statement about its preparation reflected the BANU's position that it was only necessary to demonstrate that the funds were issued legally in order to fully exonerate Stamboliiski from any wrongdoing. The Prague *Zemledelsko Zname* did not have to face any censorship, yet it virtually ignored the corruption scandal. The implications of the conception of corruption seen here that normalized the authoritarian and arbitrary manner in which Stamboliiski could operate with state funds will be discussed below. Suffice it to say here that this did not trouble the propaganda machine of the Democratic Alliance either, for their only concern was to show that Stamboliiski had appropriated the funds for his personal use.

The charges that *Zemledelsko Zname* leveled against Harlakov can be evaluated through the documentation left by the People's Court after September 9, 1944. Harlakov and his superior, the chairman of the Central Directorate of the Military League, Colonel Ivan Vulkov (general at the time of the interrogation), were questioned about their role in the June 9, 1923, coup. Their testimonies, although at times contradictory, are an invaluable source and were effectively used by Alexander Grebenarov to show the roles of Tsar Boris and the IMRO in the coup.[30] They are also the most immediate source about the found sums in Slavovitsa, even though the testimonies were recorded over two decades after the events and in a context in which downplaying culpability was advantageous. Harlakov oversaw the carrying out of the order to liquidate Stamboliiski and his statement was governed by a strategy to shift responsibility upward. He stated that the Swiss and French francs that he found were described by a commission and that he personally handed them over to General Lazarov, Colonel Z. Georgiev and others on his return to Sofia.[31] Harlakov reported that Stamboliiski had personally told him that the money had been given to him "to establish contacts with individuals abroad, in order to ease Bulgaria's situation."[32] He also claimed that the sums amounted to 18 million, but that he later found out that only 4 million reached the bank.[33]

Vulkov's statement, conversely, tended to devolve responsibility downward, although in a measured way. Being one of the chief organizers of the coup, he was able to contextualize his role in a fascinating web of relations that accommodated competing interests and disobedience and filled nearly forty pages. He related that he received the chief quartermaster, Col. Stoianov and a committee composed of some of the participants in the Slavovitsa events that brought a protocol of the found money. He remembered that the sum was around 20 to 22 million and he ordered that it be handed over to the Bulgarian National Bank. Three or four days later, Lieutenant Krustev came into his office to report that he had turned in the 3 million leva that he had received from Harlakov. Krustev's hope had been that he could receive some material assistance given his financial hardship, as a reward for the voluntary return. According to Krustev, Harlakov had also taken money. Vulkov's report avoided taking a position and only stated that Harlakov categorically denied this.[34]

The two testimonies contradicted themselves regarding how a unit of the IMRO under the leadership of Velichko (Velikoskopski) Voivoda arrived in Slavovitsa to assist in the murder of Stamboliiski. Harlakov stated that the Macedonians were sent to keep an eye on him while Vulkov argued that Harlakov had a secret agreement with Velichko and brought the Macedonians despite a prior decision by the Military League to accomplish the coup specifically without any potentially compromising connection to the IMRO. In any case, the IMRO was rewarded with between 1 and 8 million leva from the money found in Stamboliiski's villa, depending on the source one finds more credible.[35]

Given the numerous irregularities and dissatisfied with the way Harlakov had acted, the Central Directorate of the Military League discussed whether an inquiry should be started, particularly regarding the murder of Stamboliiski and the money. The decision was to have no inquiry under any circumstances because otherwise this would compromise the army's image of itself and its image before the nation and the world. Vulkov states, "In any case, [the inquiry] would have been unlikely to produce results given the mood and the regime."[36] The Ministerial Council of the Tsankov regime confirmed this decision with no deliberation. The institution of a transparent and responsible, or to say it differently an "uncorrupted" governmental ethic, was clearly not one of the regime's exigencies.

As to the reports that the money was turned over to the Bulgarian National Bank, they can only be confirmed with regards to 2 million Swiss francs. The bank did receive a letter from the Ministry of Finance on June 25, 1923, that asked it to receive from the judicial authorities, who were currently holding them, all the sums and valuables found with the former agrarian ministers.[37] The administrative council decided to convert all foreign banknotes and gold into leva and to place the sums in interest-bearing accounts. Then, on July 16, 1923, the council specifically discussed the almost 2 million Swiss francs that were confiscated from Stamboliiski.[38] The problem was that the 500 and 1,000 franc notes could not be sold in Bulgaria, and, therefore, the bank decided to deposit the sum in its Swiss bank accounts, where the money would bear interest and also increase the bank's foreign currency reserve. There is no account of any other funds to have entered the bank, which is curious given that the administrative council described the receipt, banknote by banknote, of a package from an anonymous sender that contained 98,000 Swiss francs, 60,000 French francs, 14 gold Napoleons, and 99 gold half Napoleons on April 4, 1924.[39] Protestations that Stamboliiski had not used the funds for their intended purpose notwithstanding, the fate of the confiscated money shows an even greater disregard for budgetary precision. It is conceivable that Stamboliiski could have been employing the money to buy influence and lobby in favor of reduced obligations, but it is certain that the Tsankov regime did not employ the money for reparations as intended.

The year 1923 marked the nadir of the Bulgarian monarchy in the interwar period. Not only had the BANU established majoritarian rule, but its championing of the republican idea moved in the direction of the rewriting of the constitution and drastically limiting the monarchy.[40] Conversely, 1941 was arguably its zenith, for although Boris lost some of his room to maneuver in international affairs with Bulgaria's joining of the Axis powers in March, his control of internal politics tightened, as can be attested to by the extraordinary expansion of the draconian Law for the Defense of the Nation, that same law that Tsankov had first passed in 1924 to crush agrarian and communist opposition. The less than two decades separating these dates mark not only the delegitimation and exclusion of the BANU as an autonomous force from politics but, in the aftermath of the Zveno coup of 1934 and Boris's countercoup in 1935, of all political organizations in general. The restoration of parliamentarism in the late 1930s under Boris proceeded under the continued ban on organized political life. At precisely this moment, when the memory of Stamboliiski had finally been made irrelevant, the rubberstamp XXV *Narodno Subranie* passed the "Bill for the rendering to the inheritors of the former prime minister, Alexander Stamboliiski, his former immovable property, now belonging to the state," on March 22, 1941.

## The restitution court

Dimitur Kushev, the Minister of Agriculture and State Property, appended a description of the motives behind this bill that he was sponsoring. He argued that the property in question consisted of fields amounting to about 40 decares (under 10 acres) of land in the vicinity of Sofia with two structures on them and a villa in the village of Slavovitsa. The appraisal of the fields and structures was 135,000 leva, while the villa could not be sold at auction for its starting price of 70,000 leva.[41] Due to the paltry value involved, Kushev writes, "Given that this concerns the property of a former prime minister and in order to assist his heirs and provide them with a keepsake, it would be just and beneficial if this property were turned over in ownership and possession to his heirs."[42] Parliament concurred with this evaluation and the matter of implementation was referred to the judiciary.

The willingness of the political establishment to finally resolve the outstanding questions surrounding the arbitrary and embarrassing seizure of all of Stamboliiski's property is a clear indication that the strategy of delegitimation through recourse to charges of corruption had achieved its purpose and was no longer necessary. The return of a few fields would not only enhance the moral standing of the regime but also eliminate the disparity between the hundreds of millions that Stamboliiski had supposedly stolen for his personal ends and the trivial amount of property that was in his name. That is why it is so surprising that the court that presided over the restitution adopted a maximalist interpretation of legally acquired property and initially refused to return anything that Stamboliiski had not possessed prior to the First World War. The documentation of this case, and the explication of the court's final decision, was remarkable for several reasons.[43] First, in reflecting the claims of the heirs of Stamboliiski, it provided the first justification and explanation of his possessions that defended his integrity and differentiated between the assets that were seized en masse. Second, as a consequence of the court's insistence to evaluate all the assets together, the documentation gave a very thorough evaluation of all the available information. Third, it betrayed the bias of the court given the fact that it passed the burden of proof to the heirs rather than assessing which property would have been subject to legal seizure in the first place and also due to its extremely narrow definition of legally acquired assets. The position that the court adopted, although legally sound, indicated the operation of a double standard in a politically informed corruption trial. The irony of its verdict was that as a precedent intending once and for all to bring clarity to the problem of corruption, it overreached to such an extent so as to make it inapplicable.

Stamboliiski's heirs were his widow, Milena Stamboliiska, who also represented his son, Asen Stamboliiski, still a minor, and his married daughter, Nadezhda Boiadzhieva. The protocol that they submitted to the court was complicated by the inclusion of the property of Nadezhda's father-in-law, also a plaintiff, for he too had been subject to confiscation. Nonetheless, Milena and Asen requested the return of fourteen pieces of immovable property, some household items and about 400,000 leva. Nadezhda's claim was to two parcels near Sofia and one-third ideal parts of three pieces of immovable property and one-half ideal parts of two farm plots. The court accepted that the total value of this immovable property was about 725,000 leva. However, it decided to pass judgment not only on the claims of the heirs but also on the other property that was seized, which included the banknotes that remained from the notorious four million Swiss francs. This whimsical addition was justified by the court's supposed desire for comprehensiveness.[44] In fact, however, the inclusion was made necessary by the maximalist agenda of the court.

In a paradigmatic example of doublethink, the court declared, "Given the described assets of the [suspected] of graft Alexander Stamboliiski and Grigor Boiadzhiev, according to articles 14 and 16 of the special law for confiscation of illegally acquired property, the mentioned persons, in the case of Al. Stamboliiski, his heirs, must establish the legal origin of this said property, consequently the burden of proof is on them to prove that it was acquired legally."[45] Presumed guilty unless proven innocent, Stamboliiski's heirs nonetheless complied and submitted a declaration that listed the income from Stamboliiski's fields and vineyards for the years from 1912 to 1923, his salaries, and the funds he received for travel and daily expenses. The submitted total was 4,557,814 leva, but the court accepted only 3,594,029 leva.[46]

Since the value of the requested restitution was under three-quarters of a million and that would have been affordable given Stamboliiski's earnings, the court added the villa in Slavovitsa at an inflated price. According to the testimony of the architect Georgi Lulchev, the same villa that could not be sold for 70,000 leva was valued at 1.5 million leva.[47] The court's opinion that Stamboliiski could not have saved enough to purchase all this property is questionable. However, in the process of collecting evidence, the court was able to discover an irregularity which most likely does signify malfeasance on the part of Stamboliiski. This irregularity emerged from the examination of how the 4 million Swiss francs were spent. The court established that Stamboliiski had given 100,000 Swiss francs to the secretary of the Agrarian Union, Milo Petkov; 110,000 Swiss francs, 50,000 French francs, and 42,282 dollars to Stoil Stefanov; 212,000 French francs to S. Kaluchev; 50,000 French francs and 503,000 leva to M. Borikov and the heirs of A. Frangia; but, most importantly, 309,000 Swiss francs to his in-law, Grigor Boiadzhiev. It appears that some of the money that Boiadzhiev received was used to pay off some of the mortgages Stamboliiski had taken out between 1920 and 1922.[48] The court never established how much money actually was misused; it sufficed to combine this fact with the money that was found in order to produce a verdict that only returned property that was owned prior to the First World War.

Unfortunately, this decision was not as tidy as the court would have liked it to be and it had to be revised following an appeal. First, the court had to revise upward the "legal" earnings of Stamboliiski by over one million leva. Second, the court had to modify its decision that declared all the property near Sofia to have been paid by the money given to Boiadzhiev. In fact, when forced to quantify the illegal purchase, the court found that it only affected five pieces of immovable property that were purchased between 1921 and 1922 for the sum of 391,666 leva![49] Eventually, some of the land near Sofia was returned.

As mentioned earlier, the restitution process was remarkable because its findings feigned objectivity, whereas the argumentation was sloppy and unsubstantiated. Further, in its initial decision, it stubbornly persisted in perpetuating the idea of all-encompassing corruption by Stamboliiski despite the political will to the contrary. Without intending to do so, however, the court's findings show that, at best, Stamboliiski could have misused 391,666 leva for his personal benefit. On the other hand, all his property combined, of a value slightly exceeding two million leva if the court's exaggerated valuation of the villa in Slavovitsa is accepted, is still not even half of all the money he received legally in the decade preceding 1923. This does not correspond to the image of depravity that the propaganda machine of the Democratic Alliance constructed. Further, the court records allow the researcher to perform an interesting calculation. The value of all the banknotes that were found in Stamboliiski's houses were worth slightly less than 2.5 million Swiss francs based on the June 12, 1923, exchange rates. The sums that the court found Stamboliiski to have given to various people were worth close to 850,000 Swiss francs. The missing 650,000 Swiss francs were worth about 10.8 million leva at the time of the coup, and they are remarkably close to the amount of money that is supposed to have gone to the murderers of Stamboliiski. The Ministerial Council of the Tsankov regime refused to start an inquiry that would have embarrassed them and the Military League. After all, using state money "recovered" from a corrupt prime minister to pay his assassins is not the best way to mark a turn to propriety from the "morally degenerate rule of the peasants."

Stamboliiski's corruption is laughable if confined to the few fields that he had bought; not so is the corruption of the political ethic and the procedural rules of government that his authoritarian and arbitrary disbursement of state funds shows. The way Stamboliiski managed the state, including the demand that his ministers provide him with signed letters of resignation prior to assuming their duties, indicates a governmental style that tended to erase the boundaries between the state and the political party in power. In 1923, the BANU had achieved its goal of *samostoiatelno pravitelstvo* (independent administration) based on majoritarian rule. It was poised to radically transform the social landscape of Bulgaria and the idea that BANU was a government for the people served as a justification for the taking of some liberties, as far as what could be done in the name of the people was concerned. Tsankov's Democratic Alliance could have effortlessly demonstrated this corruption, but it chose instead to focus on fantastical "colossal thefts."

The reason was that the establishment of economic democracy for the benefit of the people (the population in 1926 was still 79.76 percent agrarian) was at the heart of the program to set up a government of and for the people. The exposure of procedural corruption would not have resonated with the masses since it could easily be dismissed as a small aberration on the path to social justice. On the other hand, egregious graft could delegitimize the whole project by suggesting that state assets were being systematically plundered by a political elite under the cover of empty populist slogans. It did not matter if Stamboliiski's corruption had to be invented, so long as the peasant alternative modernity could be ended. Dimitrina Petrova has written:

> The concept of democracy in the Program of BANU contained the tendency to surpass the limits of the bourgeois democracy practiced in Bulgaria. People's rule (*narodovlastie*), towards which the union was striving, was emerging as a specific democratic regime within the bounds of the capitalist system, in which the upper bourgeoisie is limited economically and socially, and the national masses receive broad social and political rights and the opportunity to influence the administration of the country.[50]

The list of the platforms and policies of the agrarian government is too long to be included here, but a few examples can set the tone.[51] One of the first positions that the BANU adopted immediately after the First World War was to push for the confiscation of illegally acquired assets and taxation of profiteering during wartime. The law from February 21, 1919, progressively taxed the difference in profits during the period 1915–18 when compared to the baseline of 1914. Only after the BANU parliamentary delegation walked out because of the refusal by the government to enact a law for confiscations of illegal earnings or for the restitution of requisitioned livestock did the threat of a cabinet crisis force the passage of the diluted law of April 12, 1919, for confiscation. It is important to note that at the same time, the BANU was enforcing a political culture that prohibited its deputies from serving as intermediaries in business deals or from being stockholders.[52] The law for labor property from May 12, 1921, reveals the implementation of the idea that land should belong to those who worked it and as a consequence set the maximum size of family holdings to 300 decares.

## Making sense of "corruption"

What is the appropriate theoretical framework which can make the charade around Stamboliiski's "corruption scandal" intelligible? It is mostly in the past couple of decades that corruption has been theorized chiefly in the context of delegitimizing communism as a system, and it is this theoretical framework that can prove retrospectively most revealing in making sense of the accusations against Stamboliiski. A concern with corruption understood in relation to the moral degeneration of the polity or the deterioration of regimes has been a preoccupation of Western political thought from Plato and Aristotle. This classical tradition is founded upon a linkage between broad institutional decay and a subversion of the practices which underlie it—that is, the deviations from the common good. That linkage was problematized after the Second World War by the process of decolonialization and by the expansion and transformation of political science as a discipline, each in its own way. The turn toward quantification informed the functionalist work of the 1960s that focused on the ways in which corruption operated as an informal system in its own right, often with salutary effects.[53] The application of rational choice theory also resulted in market approaches that sought to examine corruption as transactions.[54] It was the sidelining of the ethical aspects of corruption in those approaches, however, that produced a reevaluation beginning in the 1990s. The reintroduction of the normative in what could be called neoclassical approaches was an effort to tackle corruption as a systemic phenomenon that seriously erodes liberal democracy.[55]

This is the theoretical conjuncture in which the present-day preoccupation with combating corruption can be situated. This preoccupation is visible at all levels, setting the public policy agenda of international bodies such as the United Nations or the European Commission, the International Monetary Fund or the World Bank, down to the nongovernmental level of Transparency International or the International Chamber of Commerce. Yet there is a fundamental disparity between the general consensus regarding the raison d'être of the crusade against corruption, the tally of the negative effects and repercussions of corruption, and the ambiguities and lack of agreement that plague efforts to establish a comprehensive and operative definition that can serve as a firm foundation for the former. Definitions which center on the transgression of the rules governing the exercise of public office for the purpose of private gain engage in a trade-off that limits the breadth in favor of precision because of their legalistic conceptualizations. Broader definitions which focus on deviant behavior that subjugates public to private interests can only provide the basis for rigorous comparisons of corruption across time and space if they engage in a positivistic reification of "public interest" into ideal types. In order to avoid this trap, this second set of definitions is forced to dilute its precision and the result is the recourse to public opinion data whenever the values or the perceptions of public interest must be specified.[56]

The tension between the quantifiable/structural and subjective/perceptual aspects of corruption tends to recede from view when corruption is politically instrumentalized. Within the hegemonic sphere that is established by the anti-corruption discourse, the imprecise boundaries of the concept do not function as a hindrance but allow enough flexibility to make corruption an ideal policy tool, particularly in the guise of scandal. In addition, the instrumentalization of corruption for political ends lends itself particularly well to the extension of vices from the personal to the systemic level. Indeed, when operational within a particular system, instrumentalized anti-corruption can play an important ethical role that calibrates social norms.

But this is not so with extra-systemic cases; the concealment of contradictions cannot be maintained when a system confronts an "other." The gap between the normative and the subjective becomes the battleground on which the power relations between these systems play out. Of interest is not the divergence of ethical norms this reveals and the resultant case for some form of relativism. Rather, the normative claims that one system imposes on another are indicative of the effort to shore up a rattled fictional coherence, to delegitimize a rival, and, most importantly, to deny that a viable alternative ever existed. It is the claim of this chapter that the obsession of the capitalist system with corruption that involves personal pecuniary enrichment is far from coincidental and is informed by the logic of capital accumulation within a free market economy. The private drive to amass, it is argued, is best defended from degenerating into corruption within the liberal market democracy. Therefore, it is imperative to delegitimize any alternative systems whose aim, at the very least, ideologically, is to excise that type of self-interest in the first place.

To better situate this chapter within the vast and multidisciplinary literature on political corruption, an examination of the major themes and approaches is warranted. For, this chapter's detailed case study is not primarily a positivistic correction to historical misrepresentation. Rather, it insists that the principles structuring its micro-historical investigations are illustrative of broader practices and operative logics which can be traced in a variety of social settings and comparative frameworks. I am interested in corruption as drama: the way charges are articulated, how they are directed, and the way they ultimately operate as a legitimizing tool in the context of systemic transition.

Unfortunately, a large portion of the social science literature on corruption is unsuitable for such an examination. Despite variations in approaches ranging from structural to interactional to functionalist, this scholarship is unified by a taxonometric methodology that strives to firmly define and classify the various types of corruption in order to then propose solutions. One end of the spectrum of these policy-centric studies is occupied by work such as Transparency International's *Rapport Mondial sur la Corruption 2004: Thème Special: la corruption politique* which reduces the definition of corruption to the "misuse of public power for private gain" and whose methodology relies on corruption perception surveys.[57] At the other end, more nuanced approaches, of which Arvind K. Jain's *The Political Economy of Corruption* is a good example, complicate the treatment of corruption through the assertion of the inseparability of economic and political analysis.[58] Nonetheless, a positivistic and moralizing component is retained.

To better grasp the assumptions underlying this type of scholarship, one can turn to the edited volume by Alan Doig and Robin Theobald, *Corruption and Democratization*.[59] It identifies the 1990s as the moment in which corruption came into the spotlight as "the most predominant issue of public concern" which, "in addition to its effects on aid, [. . .] is also widely held to deter investment, undermine good government, distorting government policy and leading to a misallocation of resources."[60] It identifies six causes for why the concern with corruption is supposed to have grown: (1) the end of Cold War caused a reexamination of great power support of manifestly corrupt regimes, (2) the undesirable consequences of the dismantling of state socialism, (3) globalization having increased competition and the threshold for corrupt transactions, (4) the revolution in information technologies and expansion of financial services, (5) the upsurge of financial scandals, and (6) the illicit drug trade.[61] A response to these causes is to be achieved through anti-corruption strategies organized around three elements: an anti-corruption agency, reform of the public sector, and promotion of strong civil society.[62] The restriction of corruption to a chain leading from postulated cause, to enactment, to curtailment, leads to essentialism with regard to the treatment of aberration through normative identity binaries concerning the notions such as the "self" or the "modern."

I reserve a more detailed commentary on this point for the conclusion of the chapter, but suffice it to say here that the central place of the end of the Cold War and the "unsatisfactory" performance of the post-socialist states owes much to the paradox that normative accounts of corruption encounter when they have to explain why the fall of the corrupt socialist system and the subsequent democratization produced more corruption rather than its opposite. It is clear, however, that the linkage between perceived corruption and post-socialism is of paramount concern to those studies that can be grouped under the rubric of the transitology of Eastern Europe.

For example, the trajectory of the work on corruption by Leslie Holmes moves from a linkage of corruption to legitimation crises that brought about the collapse of communism in *The End of Communist Power: Anti-corruption Campaigns and Legitimation Crisis* to a more developed positivistic, social science classification which relates the attempts at definition to the enumeration of causes, consequences, and measures to combat corruption in *Rotten States? Corruption, Post-Communism, and Neoliberalism*.[63] The latter study, especially, is based on the premise that corruption is a serious problem and strives to arrive at some normative criteria to address it. Other works on post-socialist states adopt the approach of indexing and reports, although policy implications are never abandoned. Two ways in which these works unfold can be seen in Daniel Smilov and Jurij Toplak's *Political Finance and Corruption in Eastern Europe: The Transition Period* and Betty Glad and Eric Shiraev's *The Russian Transformation: Political, Sociological, and Psychological Aspects*.[64]

*Political Finance and Corruption in Eastern Europe* is fascinating for the way it resists the "othering" of Eastern Europe while at the same time preserving the normative view of corruption as an impediment to progress. It does so through the revival of the currently unfashionable functionalist paradigm in the guise of defending democratization. Specifically, it warns that anti-corruption rhetoric should not be exaggerated and "should not eclipse concerns about the stability and legitimacy of party systems in transitional democracies."[65] The call for temperance is justified by the following statement: "After all, recent attempts to portray corruption as a major cause of the poor performance of democratic regimes, or even widespread poverty, are yet to be fully substantiated. On the other hand, anti-corruption drives and rhetoric have often led to the destabilization of democratic systems, or have prevented their consolidation."[66] That said, the work is a standard index of case studies that affirm the treatment of corruption as an impediment to "progress," the only caveat being that the defense of democracy has primacy over anti-corruption drives.

*The Russian Transformation: Political, Sociological, and Psychological Aspects* contains Brian Kuns's "Old Corruption in the New Russia." This chapter attempts to be nuanced by stressing that "a legalistic conception of corruption obscures a proper understanding of the power struggle currently underway" and, indeed, it refers to James Scott's treatment of corruption as the un-institutionized influence of wealth in a political system.[67] However, it is still trapped within the normative condemnation approach to corruption and ultimately relies on the legacy of socialist mentalities as the explanatory model: "the roots of corruption in the Russian Federation are to be found in the inadequacy of the Soviet command economy."[68]

In seeking to define corruption in a normative way so as to be able to generate policies for its elimination or containment, this literature begs the question of the power relations that underlie not only how corruption is defined but also how the anti-corruption discourse is instrumentalized. Contextual nuances such as Smilov's, notwithstanding, this body of scholarship is incapable of addressing the contested politics (or poetics) of corruption and its critique as a discursive practice. Said differently, "Postsocialist 'transitology' [of which corruption is a dominant trope] is, of course, the legacy of a conceptual schema wherein socialism is seen as capitalism's opposite."[69] The breaking of teleological accounts of a linear transition, of which Katherine Verdery's *What Was Socialism? What Comes Next?* is a notable example, by implication requires a reassessment of the way corruption and initiatives to counter it have been practiced in the post-1989 period.

The corrective to the above can be seen in a small but sophisticated and stimulating body of anthropological work on corruption which seeks to "interrogate the *idea* of corruption as a category of thought and organizing principle, and to examine its political and cultural implications."[70] Products of the cultural turn, these studies are concerned with the construction of meaning and processes of representation. In a counterpoint to the universalist tendencies of the structural and interactional approaches of the social sciences, Cris Shore and Dieter Haller write in *Corruption: Anthropological Perspectives*:

> To sum up, and to borrow a phrase from Levi-Strauss, corruption is "good to think with": we may not be able to unravel the paradoxes surrounding it, but we can learn a great deal about the world by interrogating the idea of corruption and exploring its many different manifestations. . . . In particular, it can help us to understand what corruption *means* in different parts of the world and how it is imbedded in everyday life; why intolerance to corruption is greater in some places than in others; how it becomes institutionalized and reproduced; and the distinctions people make between what American political scientist A.J. Heidenheimer (1989a) termed "white" corruption, "grey" corruption and "black" corruption—distinctions that go a long way towards explaining why everyday forms of corruption become accepted and institutionalized.[71]

This approach opens up a whole new set of questions and has methodological implications as well. Namely, it abandons data collection in the form of corruption perception surveys, statistics, or media coverage in favor of Geertzian "thick description" of the manifestation and reproduction of corruption in everyday life. The ethnographic approach situates corruption at the interstices of morality and power. Yet, this is far from condoning corruption through recourse to relativism:

> To put it briefly, when dealing with the complexity of corruption and abuses of power, we need to identify what aspects of the system encourage or generate illicit practices (illegal and non-illegal), what aspects could instead generate real changes and how people experience and speak about these changes. These issues point to the need to assess the precise identity of the dividing line between the legitimate and the illegitimate and of that between the legal and the moral. The next critical step lies in addressing the exact relationship of the protagonists in public life to formal law and its production and to their perceived legitimacy in the broader society.[72]

These works that examine the narratives of corruption and the politics involved are by-products of the cultural and linguistic turn. The focus on discourse as a way to get at the dynamics of knowledge production and the exercise of power can lead to insights that can serve as stimulating departure points for theoretical critique. In an extreme form, these can take the shape of Ruth Miller's *The Erotics of Corruption: Law, Scandal, and Political Perversion*, which through the teasing out of the discursive similarities between corruption scandals and pornography can expose the paradoxical "othering" of the liberal state:

> In other words, the democratic organs of the liberal state that make scandal—the revelation of corruption—possible are not actually focused on the liberal state at all, but on the "outside." They are not interested in discussing corruption "inside," nor do they (as a result) aid in the fostering of public debate or popular political access. Instead they narrate stories of external corruption and pinpoint the various ways in which those "inside" are under threat, or already infected by, these practices.[73]

I too am interested in the operational logics of the anti-corruption discourse. However, rather than aiming at a theoretical critique, I want to focus on them as political tools within very particular historical circumstances. I believe that the contribution of the historical profession to this problematic can best be made through the vehicle of micro-history, where a specific case can serve as an anchor for broader ramifications. The Stamboliiski corruption scandal has remained unexamined for nearly a century, trapped within the normative condemnations and finger-pointing of the parties involved. The consequences manifest themselves not only as hurdles to establishing a documented account of the events but also as obstacles to the examination of the sociopolitical implications of the scandal. The fact that the Stamboliiski scandal has been depoliticized by virtue of its antiquity allows a type of analysis that can decouple it from the existential implications that contemporary ones have. Via the reexamination of a historically contingent scandal from the beginning of the twentieth century, insights can be generated that are pertinent to the similarly contingent scandals at the end of the twentieth century and, by extension, about the contested understandings of corruption as cultural artifacts.

In Bulgarian history, the twentieth century was marked by two revolutionary efforts to reconfigure the structure and norms of society: first by the BANU in the 1920s and then by the Communist Party after the Second World War. The dismantling of these two systems was marked by high-profile delegitimating campaigns directed against Alexander Stamboliiski and Todor Zhivkov for corruption manifested in personal enrichment. The four-year-long *Delo #1* against Zhivkov is a well-known fiasco. Rather than trying him on which apartments he had distributed, what cars he had, or which Marxist movements he had supported, the charges of inciting ethnic hatred through the *Vuzroditelen Protses*, the renaming and expulsion of Turks and Pomaks in the 1980s, should have been pursued vigorously. Instead, by 1998 Zhivkov had died, and the trial for the *Vuzroditelen Protses* ended, while the charges that he had appropriated 26.5 million leva for himself and others over the course of twenty-seven years were a waste of time.

It should be kept in mind, however, that this was not only the case in Bulgaria. Several trials of former communist heads of state involved pecuniary corruption charges but ended with dismissal of cases or reduced prison sentences. In Poland, the impeachment proceedings against Jaruzelski in the Committee on Constitutional Responsibility in the *Sejm* reflected charges of corruption introduced by the KPN (Confederation for an Independent Poland) which produced so much public scorn that they were withdrawn.[74] Prominent among the charges that Ramiz Alia faced in Albania were abuse of power and misappropriation of state funds, yet even though he was sentenced to nine years in prison, his term was first reduced to five years, and then he was released in 1995 after having served one year.[75] The 800-page indictment against Honecker in Germany included not only the charges for collective manslaughter related to the orders for securing East Germany's border but also charges that he had stolen 9 million dollars. The trial did not take place on health grounds.[76]

A final point can be made about the centrality of pecuniary enrichment to the high-profile, grand corruption cases against communist heads of state as well as Stamboliiski. A strongpoint of the classificatory social science literature is the categorization of corruption along a multitude of practices, from bribery to clientelism, from political party finance to privatization. This is useful because the charges against Stamboliiski ignored the truly disturbing corruption of public office, the extralegal discretionary powers which were assumed by both him and his accusers, but instead focused only on whether he personally stole public funds. One of the major aims of this chapter is to illuminate the reasons for this focus because that, in turn, reveals something about the operation of corruption charges in the setting of systemic transition.

This chapter has been consistently excavating the logic of the instrumentalization of Stamboliiski's corruption in order to show its political function. After having performed the analytical work to analyze for the first time all the available evidence and to move the discussion away from polemical propaganda, its focus has shifted attention from the questions concerning what Stamboliiski stole, or if he stole at all, that are integral to the detective story of establishing the historical facts. Instead, the much broader set of questions revolving around why his "thefts" matter and how they were used have allowed the chapter to concretely illustrate the instability between the quantifiable and subjective aspects in the understanding of corruption which emerges with particular intensity when two sociopolitical systems collide.

## Radić's treason charge

In his memoirs, Vladko Maček recounts the episode of Stjepan Radić's return to the Kingdom of Serbs, Croats, and Slovenes from his trip to Moscow via Vienna on August 11, 1924.[77] Aware that an order had been given to all the border posts for the immediate arrest of Radić as soon as he would step back onto the territory of the kingdom, Maček hurried to Vienna to warn Radić. After informing the latter of the domestic situation, Maček implored him to remain a few more days in Vienna while he would return to Belgrade to resolve the matter. In Belgrade, Maček would have endeavored either to revoke the arrest warrant or, in case that was impossible, to precipitate a vote of no confidence in the *Skupština* against the Davidović government which had just been formed.[78] Radić did not agree, answering that the regime could be brought down anyway were he to be arrested, and resolved to undertake the return the following morning.

At the border, the police, having been warned from Vienna about Radić's imminent arrival, immediately sought the travelers' passports. Maček had a properly issued passport that passed a suspicious examination. Radić, on the other hand, only had an authorization issued by the London police on which were affixed various European visas. As Maček narrates, they were fortunate that a whole page was covered with the seal of the Soviet Union and Cyrillic text. Since at that time there were no diplomatic relations with the Soviet Union, the official assumed he was dealing with a special visa from Belgrade and, with the courtesy of a deep bow, allowed the voyage to continue.

Maček suspected that this was immediately reported, and a second attempt was made to rectify the official's error. A few hours into Yugoslav territory, another official demanded to see Radić's papers. Tersely, Radić retorted that he has no passport since he was travelling in his own country and that passports are only checked at the border. Maček concluded the story with, "This is how this official, too, went away without having completed his work."[79]

I narrate this story here directly as it has been recounted, because its factological utility is secondary to its revelations about the mindset of the CRPP's leadership. Whether or not the officials' inspections of their documents were orchestrated by Belgrade and overlooking the implausibility that an officer that can identify the travelers would be illiterate enough to confuse a Soviet visa with that of the Kingdom of Serbs, Croats, and Slovenes, the fear and expectation of an arrest were present in the calculations of the travelers. Counterpoised to the intimidation, one can also observe a genuine audacity that goes beyond the willingness to take risks into the hypothetical adventurism of bringing down regimes. In this seemingly innocuous anecdote, one can read the confrontational dynamics between the CRPP and Belgrade that were fueled by mistrust and paranoia.

I argue that this mindset of confrontation is the key to explaining the drama of Radić's actual arrest and subsequent trial when it came at the start of 1925 not with a whimper but a bang. Not only was Radić arrested and charged with high treason, but article XVIII of the Law for the Defense of the Realm was invoked to protect public order and the security of the state against the entire CRPP's apparatus. Wayne Vucinich writes that "the CRPP was dissolved on the ground that it had joined the Peasant International and thereby had become an integral part of the *Comintern*. Meetings and publications of the CRPP were prohibited, its archives were confiscated, and steps were taken for the prosecution of its leaders. Stjepan Radić, Vladko Maček, and others were arrested on January 5, 1925, and brought to trial."[80] The *Obznana* (notification) against the CRPP was accompanied by a delegitimation campaign in the press that included the publication of forged documents. In addition, it was marked by an inconsistent admixture of legal and extralegal maneuvers. This produced absurdities such as the fact that the top leadership's arrest contravened their parliamentary immunity. While in prison, Maček continued to regularly receive his salary as vice president of the National Assembly as well as his monthly quota of one thousand cigarettes. While he could continue to represent the *Skupština* to visitors in his prison, his signature was affixed to changes to the articles of the electoral commission without his consent.[81]

Some background is necessary to contextualize the Belgrade offensive against the CRPP. On the international front, it involved Radić's efforts to break the stalemate in Yugoslav domestic politics through an appeal to the Great Powers—first through a campaign in London to garner support for Croat republicanism and second through the ill-fated trip to Moscow. Domestically, it concerned the negotiations the CRPP led with Davidović's Democratic Party (*Demokratska Stranka*).

In London from August 17, 1923, Radić sought to gain support for the Croatian cause. He conducted a lecture tour among Labour politicians but could not translate sympathy into meaningful support. By November, all Radić could report was that his situation was getting better every day because "the Croat Question is increasingly becoming a part of the current policy of the two Great Powers [Great Britain and France] and two other such states [probably a reference to the USSR and Italy], which can help us the most."[82] On December 22, Radić left London without having secured official backing for the CRPP position. He would remain in Vienna until his voyage to Moscow on June 2, 1924, to join the *Krestintern*.

Meanwhile the political situation in the Kingdom of Serbs, Croats, and Slovenes had changed with the resignation of Pašić on April 12, 1924. Ljubomir Davidović of the Democratic Party was charged with forming a government on July 24. Having dropped the policy of abstention in March to form an opposition bloc with the Democratic Party, the Yugoslav Muslim Organization, and the Slovene People's Party, the CRPP was offered four ministerial posts. Although the CRPP was supportive of Davidović, it would only agree to enter the government on September 15, 1924, after lengthy but inconclusive negotiations. The Democratic Party was pushing the CRPP to renounce its republican and national-federative program, to restrict its activities only to parliamentary means, and to recognize the monarchy and the Vidovdan constitution. Radić was insisting on Croat autonomy, that every ministerial post have a Croatian representative attached to it, and that the Croatian *Ban* would become a vice prime minister.[83] Reviakina quotes a speech by Radić in the *Skupština* that was printed in *Izvestiia*, No.225, October 5, 1924: "They want a rupture between Moscow and me as well as with the Peasant International. . . . We were recognized abroad as representatives of the sovereign Croatian nation and I am convinced that at the moment when our existence as a nation would be threatened on the part of the Serbs, the USSR will be able to impose its political line. The same goes for England and France."[84] King Alexander was opposed to the entry of the CRPP unless it renounced its republicanism. Biondich succinctly summarizes the intrigue that followed:

> On 8 October, the king orchestrated the resignation of the Minister of Defense General Stevan Hadžić, who explained his resignation in terms of Radić's "defeatist speech" on 5 October about Yugoslav policy towards Albania. On 12 October Radić delivered a vitriolic speech attacking the Belgrade political establishment, for Hadžić's resignation was intended to undermine the Davidović cabinet and to prevent Radić's participation.[85]

When Davidović was forced to resign on October 15, the stage was set for Pašić's return in November. Due to the strong opposition, Pašić immediately dismissed the *Skupština* and called for new elections on February 8, 1925.

A Machiavellian view of the repression against Radić and the CRPP that started with the coming of 1925 would treat it as an attempt to destroy its chances for doing well in the election. Given the unprecedented levels of intimidation and the ubiquitous presence of police and soldiers that revealed a de facto state of emergency, this argument carries weight. Neither is it compromised by the strong showing of the CRPP that secured for it the second largest block in the *Skupština* after Pašić's NRS. Nonetheless, to do so is to ignore the fear and anxiety that Radić's republican program and his demands for a resolution of the Croatian question caused to the monarchy and the adherents to a unitary state in Belgrade. "It was this trip to Soviet Russia and the CRPP's accession to the *Krestintern* that ultimately provided the NRS with a suitable pretext to employ harsher measures against Radić and his republican movement."[86] The pretext, as it was articulated in the charges against Radić, was the defense of the state against revolutionary efforts to subvert and partition it with communist help. It is telling, too, that the Law for the Defense of the Realm was designed to exclude the Communist Party from the political scene in 1921 and it was now being applied against the CRPP for its alleged transformation into the arm of the *Comintern*. It was a pretext, too, because it will be shown that Radić's entry into the *Krestintern* was accomplished at his own footing and while preserving the autonomy of the CRPP. Consequently, the most satisfying characterization of the offensive against the CRPP is to treat it as a delegitimation campaign designed to weaken its organizational strength and political prestige through the misrepresented association with the bête noire of communism.

In actuality, Radić accomplished the accession of the CRPP into the *Krestintern* on his own terms, without ceding the authority and independence of his party. He agreed that the program of the *Krestintern*—the necessity for a worker-peasant bloc, the provision of land to the peasants, the right of workers to nationalize factories, the freeing of national minorities, the formation of a Yugoslav federation as well as a Balkan federation—coincided with the program of the CRPP.[87] But beyond that, in the June 27, 1924 letter request for membership to the presidium of the *Krestintern* he wrote, "[The CRPP] will preserve its program and tactics, which coincide with the chief aims of the Peasant International—with their united efforts, peasants and workers to prevent the emergence of new wars and to take power in their own hands to improve the fate of all who labor. . . . The CRPP recognizes the just demands of the workers, including the nationalization of industry. Thus, the realization of the Peasant republic in Croatia will be at the same time the realization of a Workers republic and all the aims of the working class."[88] Radić was even able to have Yugoslavia represented in the *Krestintern* not as a country, but as nations living in it. All this was possible because the *Krestintern* counted on garnering enormous legitimacy for the newly founded organization through the attraction of one of the most powerful peasant parties.

The reality of the careful and circumscribed entry of the CRPP into the *Krestintern* was irrelevant to the domestic delegitimation campaign in the Triunine Kingdom. Already on December 18, 1924, a forged document entitled "A Contract Made between the *Comintern* and the CRPP," allegedly signed by Zinoviev and Radić, was published in the newspaper of Pribičević, *Reč*. Consisting of twenty-two points, it claimed, among other things, that all propaganda and agitation by the CRPP was to be of communist character and in accordance with the resolutions of the III International; it called for the creation of parallel illegal organizations to be mobilized for class warfare, penetration, and agitation in different arenas such as the village, national minority parties, or trade unions; it required work against the League of Nations and that not only did the CRPP have to follow the directives of the *Comintern* and the precepts of communism, but it would have to change its program to accord with these.[89] The forgery was immediately challenged, even from the *Krestintern* itself. In a December 23, 1924 letter, Vice General Secretary Dombal, writing on behalf of the *Krestintern*, informed Radić that they were ready to undertake a campaign in the European press to expose the forgery and the terror, and asked Radić whether he thought that this would be useful and would suit him.[90]

The incriminations against Radić were based on scant evidence and doubtful reports as to his activities. A significant portion comprised speeches he had given, which could be defended as part of his role as a political figure. As stated, the major charge was that by joining the *Krestintern*, Radić had also joined the *Comintern* and had undertaken the path of revolution. His insistence on the separation of the *Krestintern* from the *Comintern* as only a peasant organization was met by the prosecution with efforts to find more evidence. To aid Radić, the *Krestintern* offered to send Dombal as a witness in his trial.[91]

In effect, the whole episode was blackmail. There was no communist conspiracy, no treason, no plans to foment revolution and armed struggle, or to break up the state. In the prosecution one sees the fears of Belgrade and the king, and these fears found expression in the manufactured delegitimation campaign. It is important to view the attack against Radić and the CRPP as a delegitimation campaign instead of as brute repression. Unlike the Communist Party, which had been completely banned in 1921 under the same article of the Law for the Protection of the Realm, the CRPP was permitted to run in the February elections. Given its success in the election, Pašić and King Alexander wanted to avoid making Radić a martyr. Maček reports a statement that Pašić made tête-à-tête to Radić's defense attorney, Trumbič: "I know very well that Radić is as far from communism as am I. I know very well that it was Radić himself that saved Croatia from bolshevization in the years 1919–1920 through his peasant movement. But, so, politics cannot be sentimental, so I have to use methods that can lead me to the finish."[92] The overtures by Belgrade toward Radić to resolve the conflict were reciprocated because Radić did not want to risk the destruction of his party. On March 4, 1925, the CRPP dropped "Republican" from its name and became the *Hrvatska Seljačka Stranka* (Croatian Peasant Party, CPP). Further, Stjepan Radić asked his nephew, Pavle Radić, to read the following text in the *Skupština* on March 27, with which he recognized the monarchy, the Vidovdan constitution, and the unity of the state:

> We recognize the unitary political state based on the Vidovdan Constitution with the Karadjordjevic dynasty at its head, all the while taking into account the conduct of the positive politics that are understood as the will that the Croatian people three times in a row clearly expressed at elections with regard to the political facts and the total political arrangement as it is today. The correction of these facts and this arrangement which we can approve according to our will

and awareness, has to be subject of a review of the Constitution, specifically a National Agreement between the Serbian, Croatian and the Slovenian nations. We have never fought against this state and its interests. But from the aforementioned facts that we have pointed out, resistance originated in the people itself, something that from afar was incorrectly interpreted as anti-state, although it was spontaneous and not organized by anyone. But if this situation were to continue and become more acute, especially as it was during the last election and even after the election, it could be fatal for the whole of our future. And for the sake of our future together and for the sake of the unchangeable will to deal with the Serbian people, we will do everything so that the Croatian nation instead adheres to cooperation in the national concord. But it can only be based on factual and actual equality that has to be achieved through the correction of previous policy, on the other hand concerning the above-mentioned completion of the National Agreement.[93]

With this, the way was paved for the CPP to cooperate with the politics of the National Radical Party. Days after the new government of Pašić was formed, on July 18, 1925, King Alexander amnestied Radić and the leadership of the CPP, and they were able to enter into a coalition with Pašić.

The historiographical interpretation of the resolution of the crisis of Radić and the CRPP at the start of 1925 follows several tropes. The older generation of synthetic scholarship exemplified by Joseph Rothschild in *East Central Europe between the Two World Wars* but also by Richard Crampton in *Eastern Europe in the Twentieth Century* satisfies itself with commentaries about the erratic and destabilizing behavior of Radić or statements such as "His tactics were difficult to understand and much of the confusion in the next few years followed from his maverick conduct which became 'the despair of all those Serbs who honestly desired cooperation with Croatia.'"[94] Not only the Croatian national press but also communist scholarship from 1925, due to the disappointment over Radić's abandonment of the *Krestintern*, vacillate between accounts that either focus on the repression by Pašić and the Serbs, or fault Radić for his capitulation. Hrvoje Matković in *Povijest Hrvatske Seljačke Stranke* concludes that "the price was too high (the recognition of the centralist constitution!), but it will soon be shown that it was about a temporary retreat and tactics." These analyses miss the point of the significance and implications of this manufactured delegitimation.[95]

The bolshevization charges were a straw man. It was more than clear that they were used as a pretext, but this turned them into a two-edged sword. The strong showing at the polls in the 1925 elections demonstrated that even under repression and intimidation, the CRPP could get the popular vote in Croatia.

Thus, the question of the charges against Radić and his tactical behavior needs to be seen and interpreted in light of the legitimation and delegitimation strategies employed by all sides in the first years of the Triunine Kingdom after the 1922 Vidovdan system was promulgated. Through the boycott and refusal to participate, the CRPP was testing the limits of acceptable behavior and what could be achieved. This, I contend, is the key to understanding the reasons that took Radić to London and Moscow in the first place anyway. In the symbolic capitals of the two alternative, but increasingly acknowledged, political worlds (the USSR had been recognized in 1924 by the Labour government of the British Empire and was moving out of its diplomatic isolation), Radić was trying to gauge what kind of external support he could garner for an adjustment of the domestic system of the Triunine Kingdom. Similarly, the response of the Karađorđević monarchy and Pašić was an analogous tactical maneuver to push back within the boundaries of the system. In this collusion of tactical interests, Pašić and the king were trying to prevent a trial because of the publicity, sympathy, and support that would produce for Radić, and, likewise, Radić wanted to avoid a rupture of the system that would produce the complete marginalization and destruction of the Peasant Party as a political force. The resultant "compromise" should therefore not be treated simply as a euphemism for Greater Serbian chauvinism and repression. At its core, it was illustrative of the confrontational nature of politics and the search for optimal and possible modi operandi within the present circumstances.

All of this makes the outcome—the Radić and Pašić coalition and the recognition of the State—so significant. In a sense, this episode has to be taken as a political reset, and not judged in hindsight within the teleological narrative of the unsustainability and collapse of the interwar Yugoslav project. Even more importantly, attention to this episode brings into focus the wide options of coercion and chantage that were part of the political playing field in the 1920s. The involvement of the *Krestintern* and the Soviet Union illustrate the creative involvement of the peasant movement into the major currents of European history. Likewise, the censoring of the CRPP through the charges of manufactured bolshevization attests to the fears and uncertainties of the unitarist program. The Bolshevik support for federalization (but, interestingly, not that of the CPY (Communist Party of Yugoslavia) which in 1924 diverged from Moscow on the issue of federalism) could be conflated with the autochthonous demands of the CRPP and its republican program. Thus, the delegitimation campaign against Radić is best understood as a maneuver to force a differentiation between communism and agrarianism that, at its extreme, would cause the abandonment of the republican idea and the recognition of the monarchy. The CPY was banned through the same law for the defense of the state that was being applied against the CRPP.[96] By retreating from its application, Belgrade in effect created the conditions in which the CRPP could further distinguish itself from communism. The CRPP, a political player that was too big to be ignored from accommodation within the political landscape, would be tamed by becoming structurally a party within, not without the Triunine political system. The failures of this compromise notwithstanding, this delegitimizing episode had as its effect the legitimation of the CRPP as a political party pared from its "dangerous" program and finally ready to join in the political game of the Triunine Kingdom.

Unlike in 1921 with the case of the Communist Party, the 1925 offensive against the CRPP did not follow the previous script. The party was not banned and repressed. Rather, delegitimation served as a coercive wedge to force a conceptual disassociation. Although the ties between the CRPP and Moscow were tenuous and based on the attempts of both parties to profit from the other, the possibility of an accord was threatening enough for the repressive fist of Belgrade to be raised. This fist did not fall because the soft power component, what I have been treating in terms of delegitimation, was enough to rupture the potential elaboration of a radical peasant-worker front. In removing this potentiality, delegitimation opened another: the means for an end to the politics of abstention of the Croatian agrarians and their entry into Yugoslav political life.

In contrast to the Bulgarian case, delegitimation in the Triunine Kingdom did not operate as the discrediting and supersession of a rival systemic vision. Nor did it mean the cooptation of the Croatian Peasant Party. What delegitimation actually produced was the reimagination of the party program, its horizon of possibility, and its positionality within European politics. The volte-face was in fact a tactical reset, a second chance, one of the very few that the ungenerous interwar decades would grant to the peasant moment.

It is worth emphasizing the recourse to delegitimation in contrast to the outright outlawing of parties, as was the case with the Communist Party in both countries. The CPY was the target of the *Obznana* of December 29, 1920, over deaths in a miner's strike. The *Obznana* prohibited Communist activity, resulted in the arrest of the leadership, the seizure of party property, and the closure of its press. The limits of its scope were that Communist deputies in the *Ustavotvorna Skupština* (Constitutional Assembly) could continue their work and the fifth point, which prohibited any disturbing manifestations in Belgrade during the work of the Constitutional Assembly seemed to suggest a terminal date for its validity. On July 21, 1921, less than a month after the constitution was adopted, however, the Communist terrorist cell *Crvena Pravda* succeeded in assassinating the interior minister and author of the *Obznana*, Milorad Drašković. This act, rather than compromising the *Obznana* as intended, polarized opinion against the Communist Party and resulted in the passage of the *Zakon o zaštiti javne bezbednosti i poretka u državi* (Law for the Protection of Public Security and Order in the State) which completely banned the Communist Party. Similarly, in Bulgaria, the *Zakon za zashtitata na durzhavata* (Law for the Defense of the State) was passed on January 4, 1924, as a way for Tsankov's Democratic Alliance to crush the Communist Party after the failed September Uprising of 1923, although the broad language of the law was applicable to any party promoting revolutionary struggle.

This chapter chronicled two strategies that were employed to pressure and delegitimize agrarianism. It supplemented a minute excavation of two events with a theoretical analysis of their implications. In this way, the chapter served as another demonstration of the continued relevance of the study of agrarianism. The boogeyman of the interwar capitalist order was terror and revolution and that is why the smaller and weaker communist parties in Bulgaria and Yugoslavia were outlawed while the agrarian parties were allowed to continue to operate albeit under various degrees of repression. It was the inverse with the communization of these countries after the Second World War, where oppositional political activity was disallowed and the agrarian alternative was brought to an end. If the interwar capitalist system could contain the peasant alternative, the postwar communist system eliminated it. The development of this argument will be at the core of the concluding chapter.

# 6

# Drawing the Curtain

The third chapter of this work explored the instrumentalization of nationalism in the context of agrarianism. In its latter half, it argued for the inseparability of the agrarian program from the elaboration of nationalism in either a more muted or a vocal form. That argument was framed in relation to the subaltern studies project since both intellectual exercises are motivated by the need to account for difference from a homogenizing and essentializing doxa, to perform that critique from the perspective of ignored and marginalized groups, and finally, through an archaeology of the "forgotten," to impress the indispensability of alternatives, be they on the level of discourse, practice, or in aggregate categories such as modernities. These connections between the two projects, notwithstanding, the second part of Chapter 3 also critiqued a representation of Europe by scholars such as Partha Chatterjee that denies it the very reexamination it demands for the rest of the world. The counterexample I employed was that the vibrant alternative that the agrarian political parties of Bulgaria, Yugoslavia, and Czechoslovakia elaborated between the two world wars demands space between the elimination of the peasantry in the West through the irresistibility of capitalist logic, and its eradication in the East through Stalinist collectivization and terror.

Lest it seem unfair that I single out a critically progressive body of scholarship that I admire, I want to insist that the enormous condescension of posterity toward the agrarian alternative is ubiquitous and even more reductionist in the synthetic historiography of Eastern Europe itself. Even though Richard Crampton is a specialist of Bulgaria, the chapter of his survey entitled "Ideological Currents in the Inter-War Period" feigns completeness by adding anti-Semitism to the usual suspects: communism and fascism.[1] The classic *History of the Balkans: Twentieth Century* by Barbara Jelavich skips peasant political subjectivity in favor of a brief sketch of traditional peasant life: "Although the position of the peasants has been discussed previously, the emphasis has been primarily on the great events—the wars, revolutions, and catastrophes—that changed the political status of the lands they inhabited. [*sic! the peasants as observers*]. Less attention has been devoted to those aspects of their lives that were relatively unchanging over the centuries, in particular village and family relationships. Although it is difficult to offer valid generalizations for the entire Balkans, an attempt will be made here to summarize the material conditions of peasant life and to comment briefly on family and village relationships."[2]

If political history struggles with the inclusion of social forces, then surely an economic and social historian of such stature as Ivan Berend should be able to devote more than seven pages to the peasantry in his section "Social Changes: New Forces and Factors." Yet this section from his *Decades of Crisis: Central and Eastern Europe before World War II* chiefly focuses on land reform.[3] As tempting as it is to confront this scholarship with the subject of its oversight and the weakness of its explanations for the disappearance of mass peasant politics through the simple provision of counterexamples, the result would remain unsatisfactory without an alternative explanation. While the complex social transformations that took place during the socialist period after the Second World War and again after its fall in 1989 remain outside the scope of this work, this chapter will strategically probe a few events and developments that will suggest the mechanisms through which the agrarian political project was isolated and excluded politically after the Second World War. The disappearance of agrarian politics in the countries of Eastern Europe and the disconnected anti-communist agitation of the exiled agrarian leadership in the United States meant that half of the unity that made the agrarian alternative possible was lost. The politicized agrarian actor could not survive without the party; he/she became a different actor in the encounter with state socialism. Ironically, the party leadership proved more resilient. Some of these figures agitated from abroad in the name of the millions of peasants from which they had been cut off. Their children came back to reconstitute the agrarian parties after 1989. The tragedy is not that the peasantry was doomed to extinction. The tragicomedy is that there are politicians even nowadays that struggle to speak for and exploit the memory of a peasantry that no longer exists.

It is my hope that drawing the curtain on the interwar peasant alternative modernity in this way will serve as a convincing counterpoint to the dismissive interpretations that lay the blame for failure on the supposed absence of a cohesive agrarian ideology, a disconnect between a manipulative populist leadership and the peasant masses, on overreaching radicalism, and inexperience and incapacity to govern. The golden age of the European peasantry occurred in the most inhospitable of times in which structural hurdles and political violence challenged all political currents and brought European democracy to its knees. Even faced with such difficulties, peasants articulated and defended a more equitable alternative modernity often at the cost of their lives. The challenges of reconstruction after the Second World War and the communization of what became the Soviet Block were a continuation of the prewar hurdles. Even though political action and the interwar alternative modernity had to be abandoned, it would be a disservice to the peasantry to write it off again. Recent scholarship of collectivization, arguably the most violent confrontation between the socialist state and the peasantry, reveals the inputs from below and suggests that perhaps the representation of a destruction of the peasantry under communism is less appropriate than a vision in which it participated in the construction of a new modernity.[4] Verdery and Kligman write, "We treat the collectivization process as instrumental in establishing the nature of the new Party-state itself and of its subjects. It was a defining moment both for the apparatus that initiated it and for the peasantry who suffered its consequences, with the introduction of new relations among Party cadres at various levels and between them and the rural population."[5]

Agrarian politics in the first years of the twentieth century and the first years of the twenty-first are a gulf apart. In fact, were it not for a continuity in nomenclature, there is virtually nothing agrarian in the contemporary political parties. This disconnect is evident even at the level of the party programs. For example, the Program of the Bulgarian Agrarian National Union that Alexander Stamboliiski completed while in prison during the First World War is composed of twenty-five principles. The first establishes the nature of the political entity: "The Agrarian Union, in its internal structure, in its composition, in its ideas, understandings, and goals is a totally democratic (people's rule), political-economic-social organization." Next, however, comes the all-important question of social composition: "Principle Two: The Agrarian Union is a purely agrarian organization."[6] The text of that principle is important for showing just how central the link to the social base was:

> It [the Agrarian Union] is the representative of the agrarian estate in Bulgaria. People from other estates and with oppositional interests and views to those of the agrarians cannot be in its ranks. Its principal site of action is the Bulgarian village. The natural intelligentsia of the Union is only that one, which emerges out of the agrarian estate or has immediate ties to it through a similitude of interests. Deviation from this principle is an exception, which can never and should never become a rule.... He who would ask that the doors of the Agrarian Union be opened to people of other estates, indirectly seeks the destruction of

the Union itself, for that would create a political party of the type that already exists, thanks to the lack of consciousness in the different estates, thanks to old prejudices and delusions, and thanks to the perfidy of the greedy for a parasitic life intelligentsia.[7]

The estate theory of Stamboliiski was the most radical expression in agrarianism of the rural/urban divide. The peasant parties of Yugoslavia and Czechoslovakia did not adopt quite the same form, but beneath this superficial difference lie the significant structural similarities. For Radić and Švehla, the peasant base was not just any other constituency. Nor were their parties imagined as being of the same ilk as their competitors in the political arena, with the exception perhaps of the communists. In giving expression to the peasant condition, these parties undertook an ethical intervention that had far-reaching implications for their states and societies. The outcomes differ as a result of the different national conditions, but that essential departure point, at whose heart is the peasant subject, is the crucial unifying factor.

Unlike the peasant parties of Czechoslovakia and Yugoslavia that disappeared during and immediately after the Second World War, the BANU was a significant force in Bulgaria after September 9, 1944. Even after the destruction of the BANU as an independent political force, the Left wing of the BANU had remained in the Fatherland Front and, as an ally to the Communist Party, had survived as a pro forma separate entity until 1989. However, as a meaningful political force its powers were all but curbed by the end of 1947.

After 1944, the BANU appeared as the second-most-significant political force after the communists, even as it continued to fracture. Until 1934, when all political parties were banned, the Center-Right BANU "Vrabcha 1" had been the most influential wing. But its participation in the last government before the coup of September 9, 1944, sealed its fate, and its leaders were arrested and tried by the People's Court. On the other hand, the participation of the BANU "Pladne" in the Fatherland Front secured for it the premier place in the postwar agrarian movement and participation in the government.[8]

On September 23, 1944, the leader of the BANU "Pladne," Dr. G. M. Dimitrov (Gemeto), returned to Bulgaria from emigration in Cairo to a triumphant welcome. In October 1944, he made a realistic assessment of the situation in front of the national conference of the BANU:

> First of all, we live in a state of war ... Secondly, we have the presence of Russian troops on Bulgarian soil. This in itself signifies a lot. This fact gives a new face to our social life ... And thirdly, we have a revolutionary act, such as the rupture

of 9 September. These circumstances give a fair idea about the atmosphere. Tomorrow it may change, events will unfold, but at this stage the political line of the Agrarian Union has to consider these basic circumstances.[9]

The links of G. M. Dimitrov to England, however, and his position of asserting the strength of the BANU and opposing the political claims of the communists to be the leading political force, made him an easy target, given the circumstances he was so well aware of. Taking advantage of the personal rivalries within the agrarian bloc, as well as under strong Soviet pressure and the tacit agreement of the Western allies, the communists persuaded Nikola Petkov to turn against G. M. Dimitrov. By the end of January 1925, G. M. Dimitrov had handed in his resignation from the post of chief secretary of the BANU in favor of Nikola Petkov.[10] Put under house arrest, G. M. Dimitrov managed to escape to the American embassy and, after a dramatic flight, ended his career in emigration in the United States.[11]

Nikola Petkov became the undisputed leader of the opposition, comprising the BANU as well as the much smaller Bulgarian Workers' Social Democratic Party, the group "Zveno," and the Radical Party. However, the hopes of the communists that he would be more pliant proved false. Nikola Petkov was the scion of one of the most tragic political families in Bulgarian history. His father, Dimitur Petkov (1858–1907), a mayor of Sofia, speaker of parliament, government minister and prime minister, and major leader of the Labor (later People's Liberal Party), was assassinated in the streets of Sofia. Both his sons, Petko and Nikola, became prominent leaders of the BANU. Petko Petkov (1891–1924), a close collaborator of Stamboliiski, was the director of the political department in the Ministry of Foreign Affairs in the Stamboliiski government. Arrested after the coup in 1923, he became later an MP and leader of the opposition, exposing the White Terror following the September Uprising. He was gunned down by an IMRO member who was a police employee in 1924. Nikola Petkov had been in diplomatic service in Paris and after 1923 stayed in France as a journalist. After 1929, he returned to Bulgaria and, alongside G. M. Dimitrov, headed the radical wing of the BANU "Pladne." As a member of the Fatherland Front, Nikola Petkov was an active participant in the September 9, 1944, coup. A continuous member of parliament, he became the leader of the united anti-communist opposition in the summer of 1945. Working under increasingly harsh conditions and an incessant campaign against the opposition in 1946 and early 1947, Petkov's parliamentary immunity was lifted on June 5, 1947, alongside another twenty-three deputies, the same day the US Senate ratified the peace treaty with Bulgaria, a first step toward recognizing the communist regime of Georgi Dimitrov. As Michael Boll concludes, "By now, Washington's policy of concentrating on the containment of Communism outside the borders of Eastern Europe aside from Greece and Turkey was too firmly in place to allow a major policy review for the sake of a nation in which America has experienced nothing but failures for the past three years."[12] After a trumped-up trial, Nikola Petkov was charged with espionage, and was hanged on September 23, 1947.[13]

The BANU-Nikola Petkov was dissolved in 1947 after its leader was arrested. A party that carried on his name formed in 1989, was one of the founding partners in the oppositional Union of Democratic Forces and picked up the anti-communist struggle. There was an attempt to unite the two organizations in 1991, the so-called BANU-United, but that formation could not gather the necessary votes to get parliamentary representation. Then, Anastasia Moser, the daughter of the agrarian leader G. M. Dimitrov, returned to Bulgaria from the United States and formed the conservative BANU-National Union.[14] The law for the confiscation of the property of the totalitarian organizations from 1991, which covered the Communist Party, the BANU, and the Fatherland Front, further weakened the BANU that had existed prior to 1989, and after BANU-United merged with the BANU of Moser in 1992, it ceased to exist. The BANU could not remain united. On the side of the democratic forces, it split into factions following various leaders. On the Left, already from 1993, Agrarian Union "Aleksandur Stamboliiski" separated and became a coalition partner to the Socialist Party. As has been the case in the post-socialist period, at best these parties barely met the parliamentary threshold, and their presence in parliament most often was due to the broader coalitions in which they participated.[15]

A concise way to mark the gulf separating Stamboliiski's organization from the beginning of the twentieth century from the one that bears his name at the beginning of the twenty-first is to juxtapose their programs. Of all the BANU factions that populate the Bulgarian political landscape, the one that defines itself as Center-Left and has kept the name of Stamboliiski is the Agrarian Union "Aleksandur Stamboliiski." Furthermore, from 2001 it has been a constant member of the Coalition for Bulgaria that is headed by the Socialist Party.[16] The position that it claims as a defender of social justice allows it to claim a continuity to the interwar BANU, yet its program is nothing but liberal. Its statutes from November 22, 2008, in contrast to those penned by Stamboliiski, provide status quo assurances that the party is a democratic one that organizes its activity in full compliance with the country's regulations (constitution, the Law for the Political Parties). Since its chief aims are the "the building of a democratic state, governed by the rule of law, as well as a developed civil society," Article Five declares in a hodge-podge of politically correct terminology that "the Agrarian Union 'Aleksandur Stamboliiski' will assist in the European and Transatlantic integration of the Republic of Bulgaria, in the welfare, the moral and spiritual development of the Bulgarian people and of every citizen, in the economic and social progress of the nation, for the development of agriculture and the Bulgarian village, in the full utilization of the capacities of private, cooperative, and state property, on the basis of the ecological politics of the state."[17] If the disconnect with radical agrarianism is not obvious due to the generalities in the statutes, then one only has to turn to the program of the party for confirmation:

> The contemporary global processes fatefully affect the future of Bulgaria and demand its technological, economic, social, and institutional modernization. The Agrarian Union "Aleksandur Stamboliiski" is for the policy that will lead Bulgaria out of the current crisis and will secure its modernization in accordance with the global changes and transformations in today's world. The creation in Bulgaria of a peaceful, free, democratic, and socially just society, in which the rights and the higher standard of living will be guaranteed for all Bulgarian citizens, is a core social goal of the Agrarian Union "Aleksandur Stamboliiski." This is our, *agrarian* understanding of society, which should emerge as a result of the *currently ongoing* political, economic, social, and other reforms in our country. (italics mine)[18]

The struggle to justify its existence without an agrarian base or without any significant markers to distinguish it from any other liberal party makes it imperative for the Agrarian Union "Aleksander Stamboliiski" to argue for pro forma continuity with agrarianism, even where there is none. The tension between the radical roots of the interwar BANU with its insistence on being the party of the agrarian estate and the current all-inclusive politics is not lost to the authors of the Party Program. Therefore, the conclusion of their document explicitly references the transformation of the party. The honesty of that admission, however, is more than offset by the absurdity of calling the current formation a "modern" party in opposition to what it had been before: "The Agrarian Union 'Aleksander Stamboliiski' renews itself and is transformed into a modern party, in order to be in synchronicity with the most progressive trends in European and world development. This way the Union becomes strong, vital, and capable of recreating the principles, bequeathed to us by our patron, Aleksandur Stamboliiski, in the new political conditions in our country and the world."[19]

Unlike in Bulgaria, where the numerical strength and radical roots of the agrarian movement allowed some of its currents to persevere through the Communist period, albeit in a diluted and co-opted form, the agrarian parties of Yugoslavia were quickly dissolved. In Serbia, the Left wing of the Agrarian Union had become the People's Peasant Party (*Narodna Seliačka Stranka*, NSS) led by Dragoljub Jovanović and in 1941 entered into a formal alliance with Tito, pledging support for the fight against the occupiers and for an alliance with the Soviet Union. However, at the war's end, Jovanović struggled to preserve the independence of the agrarian movement against the dictates of the Communist Party and stood up courageously in parliament in defense of basic democratic freedoms. By mid-1946, he was stripped of his parliamentary and university posts, by the end of the year he was even expelled from his own party and after a show trial in 1947 was sentenced to nine years of hard labor.[20]

The demise of the Croatian HSS started already during the war when it had to contend not only with the Ustaša but also with the communists. Vladko Maček refused twice the offer to become the prime minister of the new Independent State of Croatia, he was actively opposed to the Ustaša regime, and it sent him to the Jasenovac concentration camp and later into house arrest. At the same time, he and the party leadership were deeply distrustful of Tito's partisans and warned their cadres to "keep away from the Reds," something that left them open to accusations of collaboration with the Ustaša, Četniks and government-in-exile.[21] Maček's wait-and-see policy cost him the fracturing of his party, with many members joining the Ustaša, while others fought with the partisans. At the end of the war, Maček emigrated first to France and then to the United States.

Thus, 1947 saw the end of even pro forma political pluralism in Yugoslavia. In his speech before the Second Congress of the People's Front in September 1947, Tito summarized the situation: "All of the pre-war bourgeois parties have been discredited and have lost their right to speak on behalf of the people today. They have shown that they are incapable of running the country, that in the new social order their existence is not justified and that it [pluralism] has become superfluous."[22]

The political leadership of the agrarian movements that could flee ended up founding the International Peasant Union in New York. It included Stanisław Mikołajczyk of the Polish *Polskie Stronnictwo Ludowe "Piast,"* G. M. Dimitrov of BANU "Pladne," and Vladko Maček of the HSS. From 1950 to 1971, this organization, which presented itself as the continuation of the interwar Green International, published a *Bulletin* that informed about the transformations behind the Iron Curtain and critiqued the socialist experiment. A good way to represent the nature of this organization and to illustrate the points of divergence from the MAB is to examine the article "The Great Submerged Movements" that was written in 1949 by Vladko Maček, who was the vice president of the International Peasant Union. Ostensibly, the goal of the article, as the first paragraph states, was that "the Western World still does not know the real substance and aims of these peasant parties. I feel that this state of affair should be remedied."[23] The article laid out a genealogy of the emergence of the peasant parties in order to establish their legitimacy. Although in exile and separated from its base, the leadership of the International Peasant Union still clung to the social endorsement granted by "about two hundred million peasants [that] live on the territory behind the Iron Curtain and in the Soviet Union itself. Half of this peasant population, which before the war was outside Russia and had the opportunity to develop politically, formed its peasant parties."[24] In a departure from the rhetoric of the interwar peasant parties, however, Maček ended his article with an appeal for intervention by the West. The novel rhetoric of the article, pregnant with references to the Iron Curtain, anti-communism, and the language of rights, revealed a growing sense of isolation and hopelessness. It is important to examine this position more closely, because not that many years earlier, the peasant alternative was not a peripheral project, but a European one, and as such could appeal to itself and universals, rather than to the "West." The article concluded with what I consider to be the best self-characterization of the International Peasant Union, produced in a tone that cannot conceal its pathos:

> In the ideological struggle with Communism, peasant movements are already emerging victorious. In the struggle to overthrow the imposed Communist regimes we combined our forces and formed the International Peasant Union. As such we apply for support to free the Western democracies [*sic! should read as "the free Western democracies." Could this be an unconscious slip from the interwar years when agrarianism was held up as a more equitable system that could instruct the capitalist order?*]. The sooner and the more effective this support, the better will it serve not only our cause but also the cause of democracy and social justice in general.
>
> The Western world must realize that the peasant movements of Eastern Europe were already, between the wars, one of the strongest political factors in this most essential part of the European continent. Through the sufferings sustained during the war and nowadays, the East European peasants have become even more conscious of their human and political rights. The main obstacle to the stabilization of the present Communist regimes behind the Iron Curtain, according to the admission of their rulers, is the peasants. At present,

the peasant movements are by far the strongest political groups in Eastern Europe and the future organization of that part of Europe can be conceived only as a new order of which peasant movements are the basis and rallying-point.

Only when these facts are adequately understood can the significance of the International Peasant Union, formed in Washington, be fully appreciated.[25]

The triumphalist transitologists of the post-1989 landscape spoke of a return to diversity. With the revival of multiparty politics after 1989, peasant parties reappeared, but the disconnect that the International Peasant Union reveals is the cause for the marginal agrarian parties that did form, to generally not attempt to draw on the interwar agrarian legacy and to not adopt the names of their interwar predecessors. The exception to this is the HSS in Croatia that could draw dividends from the nationalist legacy of the interwar movement in the context of the dissolution of Yugoslavia and the creation of an independent Croatian state.

In Croatia, the HSS (Croatian Peasant Party) reformed in 1990. Occupying a centrist position, this party has been able to secure parliamentary representation even outside coalitions, although in the 2011 elections it could hold on to only one seat. Nonetheless, at its high point in 2000, as the strongest party in a coalition with, among others, the Croatian People's Party and the Liberal Party, it held seventeen seats and 11.26 percent of the popular vote. As in the case of Bulgaria's BANU, the Croatian HSS perseveres as not a completely marginal presence on the political scene, but at the cost of stale liberal politics. The language of BANU "Aleksandur Stamboliiski" that was critiqued above has a twin in the program of the HSS for the period 2012–16. The introduction to that document states, "The *HSS* as a traditional Croatian Party of the political center hereby adapts to the contemporary environment and the fact of Croatia's entry as a full member in the EU in July 2013 as one of the biggest historical challenges but also as one of the biggest developmental opportunities in modern Croatian history."[26] The document enumerates the following tasks before the party: "an exit from the long-lasting economic crisis, the ensuring of long-term sustainable economic growth, the opening of space for investment in the economy and job creation, balanced regional and *rural development* [sic], public sector reform, integration in the large EU market, and the development of society and the social services in general."[27]

Serbia saw the emergence of the NSS in 1990. This party claimed continuity from the Left wing of the interwar *Savez Zemljoradnika* under Dragoljub Jovanović. Its strongest electoral results were in the 1990 parliamentary elections, but those translated to only 1.35 percent of the vote. Subsequently it joined the short-lived Democratic Movement of Serbia for the 1993 elections. That coalition captured less that 17 percent of the vote. Under the leadership of Marijan Rističević, the party moved to the Right of the political spectrum.

In the first post-communist years in Czechoslovakia, the *Spojenectvo poľnohospodárov a vidieka* (Alliance of Farmers and the Countryside) united the two strongest parties claiming to represent agrarian interests: the *Československá strana zemědělská* (Czechoslovak Agrarian Party) and the Political Movement of Agricultural Cooperatives' Members.[28] It was not able to pass the 5 percent threshold for parliamentary representation in 1990. While there was some limited success in a new coalition with the Greens in the *Liberálně-sociální unie* (Liberal Social Union) for the 1992 elections in which the coalition took seven seats, the 1996 Czech elections were a failure and thereafter that made agrarian politics in the Czech Republic irrelevant. Of the parties that had attempted to trace links to the interwar period, only the *Republikánská strana českosloslovenského venkova* (Republican Party of the Czechoslovak Countryside) could justify this claim, partly through the continuity of old Republican members and their descendants, but it could not reach the threshold for representation in parliament.

In Slovakia, despite the renewed interest in national figures after the establishment of an independent state, among whom Milan Hodža is of particular interest and prominence in the works produced by the Slovak Academy of Sciences, the interwar agrarian party was not reconstituted.[29] The *Roľnícka strana Slovenska* (Agricultural Party of Slovakia) and the *Hnutie poľnohospodárov Slovenska* (Movement of Peasants of Slovakia) emerged in the field vacated by the Alliance of Farmers and the Countryside that had operated at the federal level in the last years of Czechoslovakia. After some small coalition successes, these two parties merged into the *Nová agrárna strana* (New Agrarian Party). That formation existed from 1997 to 1998 and then disappeared in a merger with Mečiar's Movement for a Democratic Slovakia.

\* \* \*

The communization of Eastern Europe after the Second World War is contemporaneous with the closure of the agrarian alternative. That closure, however, was facilitated, in some respects tempered, by the individual histories and trajectories of the agrarian parties, most importantly during the war years. Thus, the inclusion or exclusion of the interwar agrarian organizations in the national or popular fronts of the postwar period was the initial determining condition of the role they could play up to 1948 and, in some exceptional cases, after. Bradley Abrams, in his article "The Second World War and the East European Revolution," fleshed out the work begun by Jan Gross to complicate the presentation of the communization of Eastern Europe as having arrived on the back of Soviet tanks or as the illegitimate product of a seizure of power à la Rákosi's Salami Tactics.[30] While operating on a macro, structural level, Abrams' analysis nonetheless contributes to a problematization of the first postwar years that accommodates inputs such as the increased influence of the youth or the etatization of the economy. The result is a move that at the very least necessitates a reexamination of the period and brings to the foreground the actors and processes that were engaged in the postwar contest.

The possibility of a variance in options, instead of a clear teleology, also breathes new life in the way the historian can evaluate the national roads to socialism that generated their own internal discussions within the respective national communist parties until the Stalin-Tito split in 1948 closed the door. It needs to be stressed that this approach is not intended as an exercise in speculation over counterfactuals; from the start, even those political forces that could secure a place in the postwar order were under extreme pressure that did not significantly let up until the communist consolidation of power. However, I object to the simplistic reduction that communism eliminated agrarianism. If it is absolutely necessary to point the finger at one cause, then it was the Second World War.

The clearest case of exclusion concerning agrarianism occurred in Czechoslovakia, the country that is otherwise presented as the textbook example of the complexity of communization. It was in Czechoslovakia that the Communist Party won the first free postwar parliamentary elections on May 26, 1946, with 31 percent of the vote. The agrarian Republican Party did not partake in those elections, but the reasons for that go back to before the start of the Second World War. After the national disaster for Czechoslovakia at Munich, on September 29, 1938, first Germany annexed the Sudetenland by October 10, 1938, and then Hungary received parts of Slovakia and sub-Carpathian Ruthenia through the First Vienna Award on November 2, 1938. The president, Edvard Beneš, abdicated on October 5 and went into exile in Great Britain. Until Emil Hácha became president on November 30 and appointed the agrarian Rudolf Beran as prime minister on December 1, 1938, General Jan Syrový headed a caretaker government of experts. In the meantime, Slovakia and Ruthenia received autonomy and Czechoslovakia became a federal republic on November 23, 1938. All the political parties of the Center and Right dissolved and merged into the Party of National Unity (*Strana národní jednoty*) under the chairmanship of Beran.[31] The final issue of the *Bulletin* of the International Agrarian Bureau from December 15, 1938, contained its own obituary. Its farewell statement reads, "In these conditions, the Republican Party of Agriculturalists and Small Farmers, after having guaranteed that the legitimate interests of agriculture would be protected in the new formation, itself took the initiative to liquidate the existing political parties and to unite them in one big party, the Party of National Unity."[32]

Over the course of October 1938, the activities of the Communist Party were stopped, first in Slovakia (October 9), then in the Czech lands and Moravia (October 20) and finally in Ruthenia (October 24). The Communist parliamentarians preserved their seats until the Communist Party was banned outright and its assets were seized under Beran on December 27, 1938. The Social Democratic Party and a part of the Czech National Social Party combined into the oppositional National Party of Labor (*Národní strana práce*) on December 11, 1938. That formation survived until February 16, 1939, when it was abolished, leaving the Party of National Unity to preside alone over the "authoritarian democracy" until the end of the Second Republic on March 15, 1939.

Given this prehistory, the Republican Party was not allowed to reorganize after the Second World War; but the elimination of the agrarians as a political force happened even before the capitulation of Nazi Germany.[33] The fates of Republican cadres varied on an individual basis. Beran, for example, was sentenced to twenty years in prison. The legacy of the Republican Party—more particularly, the Green International—was revived in a series of show trials in 1952, the main one of which took place in Prague from April 23 to 26, 1952. This revival had the function of finding a scapegoat for the failures of and resistance to collectivization. The trial of Josef Kepka, Antonín Chloupek, František Topol, Otakar Čapek, Vilém Knebort, Vlastimil Klíma, Josef Kostohryz, and Václav Renč was highly publicized. A propagandistic book of almost 100 pages that went through a second edition covered the process and stoked the paranoia of capitalist encirclement and subversion. Entitled *Agents of the Green International: Enemies of Our Village*, this book printed absurd plots and confessions that amalgamated imaginary class enemies with conspiratorial international organizations. The forced confession of Renč stated, "The aim of our anti-state movement, directed by the Green International, was to destroy the independence of the Czechoslovak Republic and to connect her to the so-called European federation. According to our plan which is backed by the U.S. imperialists, all countries of the peoples' democracies would become Fascist agrarian colonies of the USA."[34]

The show trial thus reveals a combination of components that revived an antagonism directed toward an already defeated "enemy." In the context of collectivization, suddenly the kulaks (*vesničtí boháči*) became "the last bastion of reaction," a "fifth column."[35] The International Peasant Union was transformed into a tool of American imperialism, while Milan Hodža's wartime proposal for a federation in Central Europe was turned into an American version of the Third Reich's plans to convert the eastern lands into a breadbasket.

Kepka was sentenced to death; Renč received a sentence of twenty-five years of imprisonment; and the remaining defendants, life imprisonment. Other processes connected to the Green International followed in 1952, involving almost 200 people.[36] It is tempting to consider these processes as a final absurd flurry before Stalin's death and the start of de-Stalinization. However, the State Historical Archives in Sofia contain a curious folder in Fond 361k: *The Representation of BANU Abroad*. This archival unit entered the collection of the archive at an unknown date, but its 176 pages are still contained in their original folder that bears the name of the Directorate of Police, State Security (from the interwar period), except that "Police" has been scratched out and replaced with a hand-written "People's Militia" (formed on September 10, 1944).

The folder, thus, must date from the second half of the 1940s, before the point when the old stock had been exhausted and new folders had been printed. Most likely, it dates from around July 14, 1947, because that is the date that one of the figures mentioned in the file filled out an appended form for a change of address. The first sixty pages in the folder appear to be a later addition and contain a few letters and a long report by Kosta Todorov about the state of affairs in Bulgaria. What follows, however, is one of the largest caches of extant documents from the interwar Green International: the MAB. This folder is the part of Rudolf Beran's personal archive that concerns Bulgaria. It was meticulously translated, annotated, and processed by the Bulgarian interior ministry. Not only was it of interest to the Bulgarian services, but a question in Russian concerning one of the documents suggests that it was scrutinized by Soviet agents as well, perhaps offering a clue about how the archive arrived in Bulgaria.[37]

I will quote two sections from the summary report that characterizes the significance of this material, in order to show the topics of interest to the communist functionaries after the Second World War:

> The output correspondence of the International Agrarian Bureau to Beran, concerning Bulgarian questions, covers the period from 18.10.1934 to 31.8.37. It can be seen from the file that the MAB was a branch of the political intelligence

section of Beran and of the Czech agrarian party. This Bureau used its links with the Bulgarian Agrarian Union to collect information about the political situation in Bulgaria.... The political goals of the Bureau were the establishment in Bulgaria of a single-party, strong, agrarian rule and the creation of a "Great Yugoslavia." In the pursuit of these goals, the Bureau interfered in the internal affairs of the Agrarian Union by acting in favor of unification of the individual fractions of the Union.[38]

Then, with regard to the period after the Second World War:

> The correspondence reveals the tight links which existed between the Bulgarian and Czech agrarians. Some of the mentioned figures have left the ranks [излезли от строя] [sic!], for example Beran was sentenced,[39] Vulkov was hung, etc. Others, however, for example Ursin, occupy leadership positions in Czechoslovakia and extremely reactionary political positions. It is well known, that the majority of the agrarians, after the breakup of their party in the Second Czechoslovak Republic, flowed into other reactionary parties, chiefly the Democratic Party of Slovakia and the National Socialist and the Lidova in the Czech [lands], where they successfully act in leadership positions. On the other hand, the link between Bulgarian and Czech reactionaries did not end completely during the time of the occupation.... It is natural in these circumstances to expect that the links between Bulgarians and Czechs, based on the above-mentioned relations, will begin to develop anew in the soil of a common struggle against the communists.[40]

Just as in the show trials, the past was brought up to justify an offensive against an already broken "foe."

## Recapitulation

This book has been about capturing the richness of the agrarian moment in a historical setting that empowered the peasant subject and mobilized that subject toward a revision and reimagination of the social field. Although the treatment of the Bulgarian case received the most attention, because it had the momentary opportunity to implement its sweeping and uncompromising vision, this monograph has also complemented its analysis through the treatment of the more constrained Yugoslav and Czechoslovak agrarianism. Together, these three countries were the founding members of the Green International (alongside Poland) and their historical trajectories represent the three paradigmatic faces of Left of Center or Centrist agrarianism.

If this work has been about rescuing the agrarian alternative from the enormous condescension of posterity, it has avoided a number of strategies. It eschewed the mode of approaches that define and reify a static peasant subjectivity. It has refused to place itself in the position of arbiter over whether agrarian programs and praxis were viable, nor does it evaluate them on their merits. Neither does it adopt a rigid mode of classification that through a standard comparative analysis could distill either an ideal type or a series of divergences from it.

Individually, the chapters in this volume are self-contained thematic explorations of key aspects of the interwar agrarian experience. Together, they add up to an illustration of the utility of reintegrating the agrarian experiment into the interwar historiography and provide a frame for the study of agrarianism that can escape the constraints of the classificatory and ideological approaches. In addition, the demonstration of the relevance of the agrarian experience to current topics and debates is meant to serve as a further justification to why this history matters.

This monograph has recast the first half of the twentieth century as the golden age of the European peasantry. Excessive emphasis on the destruction and disappearance of the peasantry after the Second World War has tainted most scholarship with a negative fatalism concerning the inevitability of this decline. Worse, these "losers" in history have not received sufficient scholarly attention because of the assumption that they simply do not matter. In sharp contrast to this orthodoxy, I have argued that in the period between the First and Second World Wars, the peasant became a political subject, understood his plight, and organized in defense of a third road between capitalism and communism. This is why this period was a golden age; for the first time, peasant self-awareness could articulate an agrarian modernity.

The second foundational premise of this book is that without the reintegration of agrarian history, the interwar period cannot properly be understood. This premise does not derive from a positivistic claim for total inclusion but is motivated by the self-evident fact that any political actor in the interwar period, regardless of his standing on the Left or the Right of the political spectrum, had to assume a position in relation to agrarianism and the peasant. At the moment when the peasant was articulating his own subjectivity, nationalists, fascists, communists, and others were deeply engaged in articulating competing subjectivities for him. This interplay of interests around agrarianism not only makes a proper history of agrarianism indispensable for the understanding of the interwar period but also opens the door for exciting new research into the competing ideologies and political currents via agrarianism itself.

Therefore, the contribution of each chapter of this work is twofold. In the alternating series of broader synthetic work or micro-history, each chapter first provided a reassessment based either on the introduction of new source material or on the novel reading of previously examined archival sources, and, second, related its agrarian history either to its contemporary context through linkages to topics such as nationalism, the federative idea, or strategies of political delegitimation, or to contemporary historiographical issues and debates such as alterity or theories of nationalism. It is fair to say, then, that the very structure of this volume is a demonstration of why agrarianism matters, back in its heyday and now.

The agrarian movements of Bulgaria, Yugoslavia, and Czechoslovakia were forged in the revolutionary formation and transformation of these countries during the First World War. The socioeconomic and political impact of the war created the conditions that propelled agrarian political formations to the forefront of national politics. In comparison to their marginal positions prior to the war, these parties were confronted with a radically transformed sociopolitical landscape in which prewar forces and institutions were discredited, severely weakened, or even expelled. In this vacated landscape, the agrarian parties could flourish.

The experience of war also radically transformed the parties themselves. The articulation of a vision for the postwar order coupled with an enormous expansion of the parties' base meant organizational and programmatic change so that these parties could adapt to their role as a major actor, if not the principal one, on the national stage. Further, the consequence of mass politics in the demographic context of half to three-quarters of the population being engaged in agriculture produced a legitimacy and urgency in the parties based on the recognition that the time of the peasant as a subject has arrived and could not be squandered.

While the war defined the central position of the agrarian parties in the postwar period, the individual positions that they adopted and the trajectories they took produced a rich variation. The proposition was introduced that the four autonomous peasant movements that emerged out of the ashes of the First World War represented the three faces of the alternative modernity that was articulated in the golden age of the European peasantry: agrarian radicalism in Bulgaria, the peasant nation in Yugoslavia (in Croatia and Serbia), and centrist agrarianism as the guarantor of parliamentary stability in Czechoslovakia.

The projection of agrarianism was contextualized on the world stage through its organization: the International Agrarian Bureau. For reasons related to the dearth of material, to the negative evaluation from its communist-affiliated competitor, the Moscow-based International Peasant Council, better known as the Red Peasant International (*Krestintern*), to its designation as a reactionary bourgeois institution during the communist period after the Second World War that even generated show trials against its "agents," but also to the scholarly neglect that has plagued the study of European agrarianism outside certain well-trodden aspects of national history, the historiography that touches upon this organization is plagued by mystification. Conversely, this work foregrounded the potentialities and the alternative developmental streams that contextualized the founding of the International Agrarian Bureau. As a result, the history of the crystallization of the International Agrarian Bureau in 1923 was explained by multiple currents in the development and implementation of agrarianism and that two frames were useful for contextualizing that history: the federalist idea itself and the competition with the *Krestintern*. Further, the international stage, despite the hopes and efforts of the agrarian parties to recast it, was severely constrained by the Versailles system. Rather than focusing on the questions of *if* or *why* the institution was a "failure," this work insisted that the recasting of the Green International in the 1920s and 1930s can be better explained by focusing on the configurations of the international situation, as well as on the competition with the *Krestintern*.

The strongest agrarian movements of the interwar period—the agrarian movements of Bulgaria and Croatia—receive a very different treatment in the literature. Whereas the Bulgarian case has been described as anti-national(ist), the Croatian one is represented as the embodiment of a national movement. This polarity ossifies and essentializes the complex relationship of the agrarian movements to nationalism and the imagined community at a moment when the character of that community was being transformed through the introduction of the peasant subject. The different outward expressions of nationalism in the Bulgarian and Croatian agrarian movements were in fact based on very similar conceptions of the peasant rights and the peasant state. This was demonstrated through a case study of the conflict over the implementation of the BANU's orthographic reform in 1921. This micro-history was wrapped in several layers of analysis that contextualized this episode and permitted its relation to several currents in nationalism studies. The analysis traced the correspondence between the conjunctural chronology of changes in Bulgarian orthography and the mutable nature of Bulgarian nationalism by opposing this inquiry to the continuing influence of the civic/ethnic antinomy in the scholarly literature of Eastern Europe. The orthographic reform reimagined the national community and in a significant, structural way paralleled a shift in the organization of the nation. It was intended to and eventually succeeded in recasting the role of the Bulgarian citizen.

The detailed micro study of the language question culminating in the orthography debate of the 1920s showed that the spelling reform was much more than a matter of sociolinguistics. It marked a fundamental change in the Bulgarian polity and was associated with a moment of restructuring, democratization, and reshuffling of elites, the rearrangement of norms that favored different social and cultural constellations. In other words, the picture of nationalism that it elicited is one that intimately grapples with one of the most central political processes of the modern Bulgarian state, the question of democracy and popular sovereignty. Inspired from the work of the Subaltern Studies Group, theoretically, this micro study elaborated an opening in the treatment of Eastern European nationalism. While a strict transposition of concepts is counterproductive given the contextual differences, the critical juxtaposition of Eastern European agrarian nationalism to the emancipatory project of subaltern studies facilitates the extraction of the former from the stale narratives that obscure and suppress the radical agrarian merger of the nation, the peasant state, and the peasant subject.

A turning point in the book is the moment at which the aspirations of agrarianism are confronted with contextual limitations. After the coup d'état against the Stamboliiski regime in 1923, the work of the BANU was coordinated by the BANU's Representation Abroad in Czechoslovakia. It related its activities to the politics of the Republican Party that was hosting it in Czechoslovakia. In exile, the Representation Abroad represented one of the lowest points in the political fortunes of agrarianism in the interwar period. Yet, my analysis aimed to show the significance and potentialities even there. The Representation Abroad's biggest achievement was that at a time of seeming hopelessness that anything could be done to restore political normalcy and a semblance of democracy in Bulgaria, its members tenaciously kept the spark alive for themselves, the beleaguered BANU, their agrarian partners in the MAB, and in front of European public opinion.

The delegitimation campaigns against the agrarian movements in Bulgaria and Croatia offer two cases that allow the layering of neglected moments of agrarian history in theoretical analysis. The first case study examined the charges of corruption against Stamboliiski as drama: the way charges were articulated, how they were directed, and the way they ultimately operated as a legitimizing tool in the context of systemic transition. The instrumentalization of corruption for political ends lent itself particularly well to the extension of vices from the personal to the systemic level. Analyzing the growing theoretical literature on corruption, the conclusion was reached that the obsession of the capitalist system with the particular type of corruption that involves personal pecuniary enrichment is far from coincidental and is informed by the logic of capital accumulation within a free market economy. According to this logic, the private drive to amass is best protected from degenerating into corruption within the liberal market democracy. Therefore, it is imperative to delegitimize any alternative systems whose aim, at the very least, ideologically, is to excise that type of self-interest in the first place.

The second case study examined the charges of treason brought about by Radić's trip to Moscow and the entry of the CPPP into the *Krestintern*. The delegitimation campaign against Radić was designed to weaken the organizational strength and political prestige of the CPPP through the misrepresented association with the communist menace. It was essentially blackmail, a maneuver to force a differentiation between communism and agrarianism that, at its extreme, might cause the abandonment of the republican idea and the recognition of the monarchy. Terror and revolution were the monsters against which the interwar capitalist order was trying to protect itself. It succeeded in outlawing the smaller and weaker communist parties in Bulgaria and Yugoslavia but was forced to accommodate the agrarian parties, albeit under various degrees of repression.

The Second World War and the years immediately following its conclusion saw the denouement of the agrarian moment. In marking the end of the agrarian alternative, this book refused to locate it in a "tragedy" of communization. It questions the simplified model of the elimination of the peasantry as a political subject in the West through the irresistibility of capitalist logic, and in the East, through Stalinist collectivization and terror. The real tragedy, it argued, was begun in the separation of the agrarian political elite from its base during and after the Second World War and is complete now, in the post-socialist period, in the grotesque revival of agrarian parties that struggle to speak for and exploit the memory of a peasantry that no longer exists. In marking the Second World War as the revolutionary and transformative moment that took away the relevance of agrarian politics, this work complemented the treatment of the First World War as having created the conditions for its initial elaboration. Touching on recent work on collectivization, this chapter normalizes the socialist experience and proposes that the peasantry, having already constituted itself as a political subject in the interwar period, was transformed as much as it transformed itself in the encounter with state socialism.

This work offers an alternative definition of agrarianism that is predicated on three universal initiatives that underwrite the agrarian project: parliamentarism, land reform, and the cooperative movement. In this way, it avoids the pitfalls of classificatory definitions and an overemphasis on ideology. While the conceptual breadth and ambition of this book do not allow it to flesh out many aspects of interwar agrarianism, the framework that it provides can serve as the foundation for further study. The nuanced framework that it proposes hopes to breathe new life into this subject that could finally allow agrarianism to receive its proper due as one of the most original and significant political currents of twentieth-century European history.

# Notes

## Introduction

1. https://www.gettyimages.ca/detail/news-photo/the-spectre-of-famine-approaches-the-well-fed-aleksandar-news-photo/526931924.
2. I was provided with a link to the Wikipedia page of Michael Nicholson. https://en.wikipedia.org/wiki/Michael_Nicholson.
3. See, for details, http://www.fundinguniverse.com/company-histories/corbis-corporation-history/, which it based on *International Directory of Company Histories*, vol. 31. St. James Press, 2000.
4. Aleksandur Bozhinov. *Album ot karikaturi iz nashiia obshtestven i politicheski zhivot.* Sofia: Pechatnitsa V. Poshta, 1907; Aleksandur Bozhinov. *Karikaturi.* Sofia: Izdava komiteta za chestvuvane 25-godishnata deinost na hudozhnika, 1924; Aleksandur Bozhinov. *Karikaturi i skitsi.* Sofia: Izdanie na Bulgarskata Akademia na naukite, 1957; Atanas Stoikov. *Bulgarskata karikatura: kratuk ocherk za neiniia put i za tvorchestvoto na Aleksandur Bozhinov, Iliia Beshkov, Aleksandur Zhendov, Boris Angelushev i Stoian Venev.* Sofia: Bulgarski Hudozhnik, 1970; Aneliia Nikolaeva, and Ruzha Marinska, eds. *Aleksandur Bozhinov, 1878–1968.* Sofia: Natsionalna Hudozhestvena Galeriia, 1999; Aleksandur Bozhinov. *Za pruv put Aleksandur Bozhinov razkazva.* Sofia: Iztok-Zapad, 2015; Aleksandur Bozhinov. *Minali dni, izdanie dopulneno s nepublikuvani snimki.* Sofia: Iztok-Zapad, 2017. It is possible that the original is in the personal archive of Bozhinov at the BAS (Fond 120k).
5. Bozhinov. *Za pruv put Aleksandur Bozhinov razkazva.*
6. Elinka Boiadzhieva. "Zhultite paveta—keramika, estetika, simvolika." *Nauka* 21, no. 5 (2011), 64–67.
7. See, for example, Alexander V. Chayanov. *The Theory of the Peasant Economy*, D. Thorner, B. Kerblay, and R. E. F. Smith, ed. Homewood, IL: Richard D Irwin Inc for the American Economic Association Translation Series, 1966; James C. Scott. *The Moral Economy of the Peasant: Rebellion and Subsistence in Southeast Asia.* New Haven, CT: Yale University Press, 1976; Theodore Shanin, ed. *Peasants and Peasant Societies.* Harmondsworth: Penguin, 1971; Erik Wolf. *Peasants.* Englewood Cliffs, NJ: Prentice-Hall, 1966.
8. *Tsentralen Durzhaven Arhiv* (Bulgarian Central State Archives, TsDA), Fond 361k, opis 1, a.e.9, f. (folio) 7.

9. Luiza Reviakina. *Kominternut i Selskite Partii na Balkanite: 1923–1931*. Sofia: Akademichno Izdatelstvo "Prof. M. Drinov," 2003; Tsenko Barev. *Prinos kum istoriiata na Bulgarskiia Zemedelski Naroden Suiuz*. Sofia: Bulvest 2000, 1994; Dimitrina Petrova. *Bulgarskiiat Zemedelski Naroden Suiuz, 1899–1944*. Sofia: Fondatsiia "Detelina," 1999; Dimitrina Petrova. *Samostoiatelnoto upravlenie na BZNS, 1920–1923*. Sofia: Durzhavno Izdatelstvo Nauka i Izkustvo, 1988; Dimitrina Petrova. *Aleksandur Stamboliiski—durzhavnikut reformator*. Stara Zagora: Izdatelstvo "Znanie," 1995; Nikolai Poppetrov. *Fashizmut v Bulgaria. Razvitie i proiavi*. Sofia: Kama, 2008; Nikolai Poppetrov. *Sotsialno naliavo, natsionalizmut—napred: Programni I organizatsionni dokumenti na bulgarski avtoritaristki natsionalisticheski formatsii*. Sofia: IK Gutenberg, 2009; Tsocho Biliarski, ed. *BZNS, Aleksandur Stamboliiski i VMRO: Nepoznatata voina*. Sofia: Aniko, 2009.
10. Momčilo Isić. *Seljaštvo u Srbiji 1918–1941*. Belgrad: Institut za noviju istoriju Srbije, 1995; Nadežda Jovanović. *Zemljoradnička levica u Srbiji 1927–1939*. Belgrade: INIS, 1994; Milan Gaković. *Savez Zemljoradnika (Zemljoradnička Stranka 1919–1941)*. Priredio Zdravko Antonić. Banja Luka: Akademija Nauka i Umjetnosti Republike Srpske, 2008.
11. Ljubo Boban. *Maček i politika Hrvatske seljačke stranke 1928–1941*. Zagreb: Liber, 1974; Bogdan Krizman, *Korespondencija Stjepana Radića, 1919–1928*. Zagreb: Institut za hrvatsku povijest, 1973; Fikreta Jelić-Butić. *Ustaše i Nezavisna Država Hrvatska 1941–1945*. Zagreb: Liber, 1977; Branka Boban. *Demokratski nacionalizam Stjepana Radića*. Zagreb: Zavod za hvtsku povijest Filozofskog fakulteta Sveučilišta u Zagrebu, 1998; Hrvoje Matković. *Povijest Hrvatske Seljačke Stranke*. Zagreb: Naklada Pavičić, 1999; Zdenka Radelić. *Hrvatska Seljačka Stranka (1941–1950)*. Zagreb: Hrvatski institut za povijest, 1996; Tihomir Cipek. *Ideja hrvtske države u političkoj misli Stjepana Radića*. Zagreb: Alenea, 2001.
12. Miroslav Peknik, ed. *Milan Hodža: Statesman and Politician*. Bratislava: VEDA Publishing House of the Slovak Academy of Sciences, 2007; Miroslav Peknik, ed. *Milan Hodža: politik a žurnalista*. Bratislava: VEDA, Vydateľstvo SAV, 2008; Pavol Lukáč. *Milan Hodža v zápase o budúcnosť strednej Európy v rokoch 1939–1944*, Štefan Šebesta, ed. Bratislava: VEDA, Vydateľstvo SAV, 2005.
13. Vladimir Dostál. *Agrární strana. Její rozmach a zánik*. Brno: Atlantis, 1998; Vladimir Dostál. *Antonín Švehla. Profil československého státníka*. Praha: Státní Zemedelské Nakl., 1990; Eduard Kubů, ed. *Mýtus a realita hospodárské vyspelosti Ceskoslovenska mezi svetovými válkami*. Praha: Nakl. Karolinum, 2000; Jiří Šouša. "K vývoji ćeského zemědělství na rozhraní 19. století. Česká zemědělská rada 1891–1914," in *Acta Universitatis Carolinae Philosophica et Historica*, vol. XCVII. Prague: Universita Karlova, 1986; Jiří Šouša. "Zemědělská správa a některá východiska vzestupu agrární strany (1914-1918)," in *Acta Universitatis Carolinae Philosophica et Historica*, vol. IV–V. Prague: Universita Karlova, 1987.

14 John Bell. *Peasants in Power: Alexander Stamboliiski and the Bulgarian Agrarian National Union 1899-1923*. Princeton, NJ: Princeton University Press, 1977; Mark Biondich. *Stjepan Radić, the Croat Peasant Party, and the Politics of Mass Mobilization, 1904-1928*. Toronto: University of Toronto Press, 2000; Daniel E. Miller. *Forging Political Compromise: Antonín Švehla and the Czechoslovak Republican Party 1918-1933*. Pittsburgh, PA: University of Pittsburgh Press, 1999.
15 Heinz Gollwitzer, ed. *Europäische Bauernparteien im 20. Jahrhundert*. Stuttgart: Gustav Fischer Verlag, 1977.
16 Eduard Kubů, Torsten Lorenz, Uwe Müller, and Jiří Šouša, eds. *Agrarismus und Agrareliten in Ostmitteleuropa*. Berlin-Praha: Berliner Wissenschaftsverlag—Dokořán, 2013.
17 Ghita Ionescu and Ernest Gellner. *Populism: Its Meaning and National Characteristics*. New York: Macmillan, 1969.
18 Louis Althusser. "Ideology and Ideological State Apparatus (Notes towards an Investigation)," in Louis Althusser, *Lenin and Philosophy and Other Essays*. New York: Monthly Review Press, 2001, 85–132.
19 M. M. Goranovich. *Krah zelenogo internatsionala*. Moskva: Nauka, 1967.
20 George D. Jackson, Jr. *Comintern and Peasant in East Europe*. New York: Columbia University Press, 1966.

# Chapter 1

1 *Zemledelsko Zname*. Br. 74. March 21, 1923.
2 "The HRSS had 'united the entire Croat peasantry and the majority of the working class, the bourgeoisie, and intelligentsia, *therefore the entire Croat people*,'" Biondich. *Stjepan Radić*, 189, quoting a declaration, *Preko pola miljuna naroda na velikim skupštinama HRSS*, Historical Archive of Zagreb, HSS, Box 14: no. 47 (1923).
3 Paul Fussel. *The Great War and Modern Memory*. New York: Oxford University Press, 2000; Jay Winter. *Remembering War: The Great War between History and Memory in the 20th Century*. New Haven, CT: Yale University Press, 2006; Jay Winter and Blaine Baggett. *The Great War and the Shaping of the 20th Century*. New York: Penguin, 1996; Jay Winter. *Sites of Memory, Sites of Mourning: The Great War in European Cultural History*. Cambridge: Cambridge University Press, 1998; Jay Winter. *The Legacy of the Great War: Ninety Years On*. Columbia: University of Missouri Press, 2009; Annette Becker, Stéphane Audoin-Rouzeau, and Leonard Smith, eds. *France and the Great War*. Cambridge: Cambridge University Press, 2003; Maria Bucur. *Heroes and Victims: Remembering War in Twentieth Century Romania*. Bloomington: Indiana University Press, 2009; Jason Crouthamel. *The Great War and German Memory: Society, Politics and Psychological Trauma. 1914-1945*. Liverpool: University

of Liverpool Press, 2010; Karen Petrone. *The Great War in Russian Memory*. Bloomington: Indiana University Press, 2011.

4 Fritz Fellner. "The Dissolution of the Habsburg Monarchy and Its Significance for the New Order in Central Europe: A Reappraisal." *Austrian History Yearbook* 4 (1968), 4.

5 Ibid., 18–19.

6 Charles Maier. *Recasting Bourgeois Europe*. Princeton, NJ: Princeton University Press, 1975.

7 Bell. *Peasants in Power*, 122.

8 Ibid., 123.

9 *Zemledelsko Zname*, N. 69 and 74, 8, August 27, 1923, quoted in Bell. *Peasants in Power*, 104.

10 Nikola Petkov. *Aleksandur Stamboliiski: lichnost i idei*. Sofia: Pechatnitsa "S.T. Charukchiev," 1930, 1946. Petkov's rendition is interesting for his commentary, but the full texts of the speeches and conversations between Stamboliiski and Tsar Ferdinand can be found in Nencho Dimov, Iordan Zarchev, and Bogomil Volov, eds. *Aleksandur Stamboliiski: izbrani proizvedeniia*. Sofia: Izdatelstvo na BZNS, 1979, 192–217.

11 Petkov erroneously gives the date as August 20, 1915. See also Bell. *Peasants in Power*, 118–21.

12 Bell. *Peasants in Power*, 110.

13 Dimov, Zarchev, and Volov, eds. *Aleksandur Stamboliiski: izbrani proizvedeniia*, 197.

14 Ibid., 198.

15 Ibid., 199, emphases in original.

16 Ibid., 202.

17 Tsenko Barev succinctly sums up the periodization: "On the next day, 11 September, the rumor is spread that Aleksandur Stamboliiski is dead so that possible national unrest could be prevented *in utero*. On the 12th of September an attempt is made on the life of the agrarian leader, and on the 13th of September an order for his arrest is issued, and on the 15th of the same month he is already in the cells of the Radoslavov police." Barev. *Prinos kum istoriiata na Bulgarskiia Zemedelski Naroden Suiuz*, 111–12.

18 Petkov. *Aleksandur Stamboliiski*, 15.

19 Kosta Todorov. *Balkan Firebrand: The Autobiography of a Rebel, Soldier and Statesman*. Chicago, IL: Ziff-Davis, 1943, 92.

20 Bell. *Peasants in Power*, 128.

21 Alexander Rabinowitch. *The Bolsheviks Come to Power: The Revolution of 1917 in Petrograd*. Chicago, IL: Haymarket Books, 2004, 48–50.

22 Nencho Dimov, Iordan Zarchev, and Bogomil Volov, eds. *Aleksandur Stamboliiski: izbrani proizvedeniia*. 2nd ed. Sofia: Izdatelstvo na BZNS, 1980, 155–79.

23  Ibid., 210. Regarding this conversation, which appears in numerous editions of Stamboliiski's work, John Bell writes, "Stamboliski published this account immediately after the meeting. His description of what took place and what was said was never challenged by the others who were present." Bell. *Peasants in Power.* Footnote 100, 120.
24  Dimov, Zarchev, and Volov, eds. *Aleksandur Stamboliiski: izbrani proizvedeniia*, 214.
25  Ibid., 217.
26  For the history of the uprising and the differing evaluations, see Hristo Hristov. *Revoliutsionnata kriza v Bulgariia prez 1918-1919*. Sofia: Izdatelstvo na Bulgarskata Komunisticheska Partiia, 1957. Dimitur Dimov. *BZNS i voinishkoto vustanie*. Sofia: Izdatelstvo na BZNS, 1979. Liubomir Ognianov. *Voinishkoto vustanie: 1918*. Sofia: Nauka i Izkustvo, 1978. For the general context of the collapse of the front, see Richard C. Hall. *Balkan Breakthrough: The Battle of Dobro Pole, 1918*. Bloomington: Indiana University Press, 2010.
27  Bell. *Peasants in Power*, 132.
28  Svetla Daskalova and Dimitur Tishev, eds. *Raiko Daskalov: Izbrani proizvedeniia*. Sofia: Izdatelstvo na BZNS, 1986, 170–85.
29  Ibid., 185.
30  Todorov. *Balkan Firebrand*, 103.
31  Bell. *Peasants in Power*, 137–38. Joseph Rothschild. *The Communist Party of Bulgaria: Origins and Development, 1883-1936*. New York: Columbia University Press, 1959.
32  Petrova. *Samostoiatelnoto upravlenie*, 12.
33  Bell. *Peasants in Power*, 142–46; Petrova. *Samostoiatelnoto upravlenie*, 34–37.
34  Augusta Dimou. *Entangled Paths toward Modernity: Contextualizing Socialism and Nationalism in the Balkans*. Budapest: Central European University Press, 2009, 96.
35  Ibid., 220–21.
36  Gaković. *Savez Zemljoradnika (Zemljoradnička Stranka)*, 30; Vladimir Dedjer, Ivan Božić, Sima Ćirković, and Milorad Ekmečić. *History of Yugoslavia*. New York: McGraw-Hill, 1975, 501.
37  Friedrich Wilhelm Raiffeisen was first involved in the setting up of charitable cooperatives to help peasants in the mid-nineteenth century and then moved to the creation of a banking cooperative in Anhausen in 1862. The principle of providing banking and credit services to the rural population made this model attractive for emulation.
38  For the early history of the cooperative movement, see Mihailo Avramović. *Trideset godina zadružnoga rada 1894-1924*. Belgrade: Zemunska štamparija glavnoga saveza srpskih zemljoradničkih zadruga, 1924.
39  Biondich. *Stjepan Radić*, 89.

40. Ibid., 160.
41. Ivo Banac. *The National Question in Yugoslavia: Origins, History, Politics*. Ithaca, NY: Cornell University Press, 1984, 237.
42. Miller. *Forging Political Compromise*, 48.
43. Ibid., 50. With the Czech Populists, these parties constituted the Pětka.
44. Ibid., 75.
45. Ibid., 33–34.
46. Biondich. *Stjepan Radić*, 95.
47. Bulgarian Central State Archives, Fond 361k, opis 1, a.e. 9, f. 7.

# Chapter 2

1. Walter Littlefield. "The Little Anti-Red Entente." *New York Times* (April 10, 1921), 88. I have purposefully preserved the multiple cases of the misspelling of proper names in the article.
2. Ibid.
3. Ibid.
4. Ibid.
5. *Zemledelsko Zname*. No 47, March 6, 1918.
6. Народно Събрание. Стенографски дневници, XIX ОНС, III Р. С., София, 1921, 132–38. Reprinted in Dimov, Zarchev, and Vulov, eds. *Aleksandur Stamboliiski: izbrani proizvedeniia*, 388
7. Ibid.
8. For the function of the social imaginary, see Cornelius Castoriadis. *The Imaginary Institution of Society* [*L'Institution imaginaire de la société*]. Cambridge, MA: MIT Press, 1987.
9. ЦДА (Централен Държавен Архив), (Bulgarian Central State Arhives, TsDA), Fond 176K, opis 4, a.e. 2026, f. 41.
10. Arhiv Jugoslavije (Archives of Jugoslavia, AJ) Fond 80, f. 80, not paginated.
11. Arhiv Jugoslavije Fond 80, f.80, not paginated.
12. Ibid.
13. Genovski describes this meeting in his book *Aleksandur Stamboliiski: otblizo i otdaleko*. Sofia: Izdatelstvo na BZNS, 1982. The narrative of this meeting is based on his archived and documented memoirs "Sdruzhenite zemedeltsi v Edinniia front." Where possible, he has complemented his memoirs with archival material such as Stamboliiski's pocket diary.
14. Genovski. *Aleksandur Stamboliiski: otblizo i otdaleko*, 101.
15. Ibid., 96.
16. Biliarski, ed. *BZNS, Aleksandur Stamboliiski i VMRO*, 111.

17 Народно Събрание. Стенографски дневници, XVIII ОНС, I Р. С., София, 1919, 516–17. Reprinted in Dimov, Zarchev, and Vulov, eds. *Aleksandur Stamboliiski: izbrani proizvedeniia*, 464–65.
18 Ibid., 466.
19 Biliarski, ed. *BZNS, Aleksandur Stamboliiski i VMRO*, 121.
20 The reparations burden was 2.25 billion gold franks, to be paid over the course of thirty-seven years at 5 percent interest.
21 For a detailed treatment of the reparations question, see Panaiot Panaiotov, "Borbata na samostoiatelnoto pravitelstvo na BZNS za po-spravedlivo razreshavane na materialnite i finansovite sanktsii na Nioiskiia miren dogovor (1920–1923)" in Konstantin Kosev, Vitka Toshkova, and Panaiot Panaiotov, eds. *Problemut za mira v mezhdunarodnite otnosheniia 1919–1980*. Sofia: Izdatelstvo na Bulgarskata Akademiia na Naukite, 1987, 7–63.
22 Bell. *Peasants in Power*, 194.
23 Todorov. *Balkan Firebrand*, 152, 163, 169.
24 Bell. *Peasants in Power*, 205.
25 AMZV (Archive of the Foreign Ministry) Praha, II sekce, kart. 427, č.j. 3111/20; III sekce, kart. 910, č.j. 6249/20. Reprinted in *Izvori za bulgarskata istoriia*, t. XXIII: Chehoslovashki izvori za bulgarskata istoriia, t.I, ed. Jozef Kolarzh, Ivan Shtovichek, Valerian Bistritski, Vasil Vasilev, Linda Manolova. Sofia: Izdatelstvo na BAN, 1985, 72.
26 AMZV Praha, PZ Brusel 1920, č. 143/20, in *Izvori za bulgarskata istoriia*, t. XXIII, 66.
27 AMVZ Praha, PZ Bělehrad 1920, č. 452/1920; TO 4021-4030 in *Izvori za bulgarskata istoriia*, t. XXIII, 69–70.
28 ЦДА (Централен Държавен Архив), Fond 176K, opis 4, a.e. 1380, f. 186.
29 Ibid., 174.
30 KPR (Chancellery of the President of the Republic)- D 9706/20 in *Izvori za bulgarskata istoriia*, t. XXIII, 80.
31 Ibid., 88. Ernest Hemingway, in his coverage of the Genoa Conference described Stamboliiski as "chunky, red-brown-faced, has a black mustache that turns up like a sergeant major's, understands not a word of any language except Bulgarian, once made a speech of fifteen hours' duration in that guttural tongue, and is the strongest premier in Europe—bar none." Quoted in Saturnino M. Borras, Jr., Marc Edelman, and Cristóbal Kay, eds. *Transnational Agrarian Movements Confronting Globalization*. Malden, MA: Wiley-Blackwell, 2008, 5.
32 Feliks Gross, ed. *European Ideologies: A Survey of 20th Century Political Ideas*. Freeport, NY: Books for Libraries Press, 1948, 955.
33 Leften S. Stavrianos. *Balkan Federation: A History of the Movement toward Balkan Unity in Modern Times*. Hamden, CT: Archon Books, 1964, 66–63; Alina Teslaru-Born. *Ideen und Projekte zur Föderalisierung des Habsburgischen Reiches mit*

*besonderer Berücksichtiging Siebenbürgens 1848–1918*. Frankfurt-am-Main: Johann-Wolfgang-Goethe-Universität zu Frankfurt am Main, 2005.

34 Varban Todorov. *Greek Federalism during the Nineteenth Century (Ideas and Projects)*. Boulder, CO: East European Monographs, N. CDVIII, 1994, viii–ix.
35 Cited in Leften S. Stavrianos. "The Balkan Federation Movement: A Neglected Aspect." *The American Historical Review* 48, no. 1 (October 1942), 32–33.
36 Ibid., 37.
37 Ibid., 44.
38 Noel Buxton. "The Balkans Today." *Nineteenth Century* XC (August 1921), 333–35, cited in Stavrianos. "The Balkan Federation Movement," 46.
39 *La fédération balkanique*, July 15, 1924, cited in Stavrianos. "The Balkan Federation Movement," 46.
40 *La fédération balkanique*, October 1, 1924, cited in Stavrianos. "The Balkan Federation Movement," 46.
41 For the detailed analysis of this episode, see Chapter 5.
42 Stavrianos. "The Balkan Federation Movement," 46.
43 The most detailed account of the Balkan conferences remains Theodore Geshkoff. *The Balkan Union: A Road to Peace in Southeastern Europe*. New York: Columbia University Press, 1940. See also Robert Joseph Kerner and Harry Nocholas Howard. *The Balkan Conferences and the Balkan Entente 1930–1935: A Study in Recent History of the Balkans and the Near Eastern Peoples*. Westport, CT: Greenwood Press, 1936 and Stavrianos. *Balkan Federation*, 224–58. Pavlos Hatzopoulos. *The Balkans beyond Nationalism and Identity: International Relations and Ideology*. London: Tauris, 2008 identifies three challenges to nationalist ideology in the interwar years: the communist, the agrarian, and the liberal internationalist.
44 ЦДА (Централен Държавен Архив), Fond 361K, opis 1, a.e. 87, f. 144–146.
45 Milan Hodža. *Federation in Central Europe: Reflections and Reminiscences*. London: Jarrolds Publishers, 1942.
46 Ibid., 8.
47 Ibid.
48 Ibid., 5–6.
49 Reviakina. *Kominternut*, 10. To a great extent, the following survey and analysis of the *Krestintern* follows the magisterial work of Reviakina, who took advantage of the declassified former Soviet archives to offer the definitive work on the *Comintern* and the agrarian parties. Reviakina exceeded all measures of generosity by sharing her research notebooks of her own abortive foray into the turbulent waters of the Green International with me.
50 *Pravda*, N.134, June 19, 1923; Reviakina. *Kominternut*, 23–24.
51 Reviakina. *Kominternut*, 30.
52 Jackson, Jr. *Comintern and Peasant in East Europe*, 68–71; Reviakina. *Kominternut*, 26–29.

53 Reviakina. *Kominternut*, 41–45.
54 *Kominternut v Bulgaria*. T.1 Sust. Luiza Reviakina, Luchezar Stoianov, Volodia Milachkov, R. P. Grishina, and N. S. Lebedeva, *Arhivite Govoriat*, t.36. Sofia: Glavno upravlenie na arhivite pri ministrskia suvet, 2005, 92–93, Document N.23 (July 21, 1923).
55 Reviakina. *Kominternut*, 48–49.
56 Ibid., 193–97.
57 G. Livingstone. "Stepan Radic and the Croatian Peasant Party. 1904 – 1929." Unpublished PhD dissertation, Department of History, Harvard University, 1959, 462.
58 Reviakina. *Kominternut*, 87–88.
59 Vladko Maček. *In the Struggle for Freedom*. New York: Pennsylvania State University Press, 1957, 54.
60 Biondich. *Stjepan Radić*, 198.
61 Reviakina. *Kominternut*, 96–97.
62 Ibid., 238–39.
63 Ibid., 231–35.
64 Asher Hobson. *The International Institute of Agriculture; an Historical and Critical Analysis of Its Organization, Activities and Policies of Administration*. Berkeley: University of California Press, 1931. Olivia Rossetti Agresti. *David Lubin: A Study in Practical Idealism*. Berkeley: University of California Press, 1941. E. Kubů and J. Šouša. *Die Wiener Grüne Internationale—eine mitteleuropäische Transfergeschichte?* in Helga Schultz and Angela Harre, eds. *Bauerngesellschaften auf dem Weg in die Moderne. Agrarismus in Ostmitteleuropa 1880 bis 1960*. Wiesbaden: Harrassowitz, 2010, 243–56.

# Chapter 3

1 Banac. *The National Question in Yugoslavia*. For a rare, alternative, recent framework of nationalism in Yugoslavia, see Max Bergholz. *Violence as a Generative Force: Identity, Nationalism, and Memory in a Balkan Community*. Ithaca, NY: Cornell University Press, 2016.
2 See, for example, Peter Sugar and Ivo Lederer, eds. *Nationalism in Eastern Europe*. Seattle: University of Washington Press, 1969.
3 This pejorative term designates the traitorous betrayal by "ethnic" Bulgarians of their "true" identity in favor of a Serb one, in this case. It was first applied toward Bulgarians identifying as Greeks or speaking Greek rather than their "mother tongue" during the Bulgarian National Revival—*Gurkoman*.
4 Spas Raikin. *Politicheski problemi pred bulgarskata obshtestvenost v chuzhbina: sbornik statii, eseta, studii i komentari po politicheski, kulturni, ikonomicheski i*

*istoricheski vŭprosi, publikuvani v emigratsia 1978–1991*, vol. 1. Veliko Turnovo: Pechat DF "Abagar."1993, 175.

5. Dimov, Zarchev, and Volov, eds. *Aleksandur Stamboliiski: izbrani proizvedeniia*, 206.
6. Bell. *Peasants in Power*, 183, 207.
7. *Zemledelsko Zname*. November 5, 1919.
8. Nira Yuval-Davis. *Gender and Nation*. London: Sage, 1997.
9. Louis Snyder. *The Meaning of Nationalism*. New Brunswick, NJ: Rutgers University Press, 1954.
10. Carleton Hayes. *The Historical Evolution of Modern Nationalism*. New York: MacMillan, 1931, 1948.
11. Hans Kohn. *Prophets and Peoples: Studies in Nineteenth Century Nationalism*. New York: Macmillan, 1946.
12. Tom Nairn. *The Break-up of Britain: Crisis and Neo-Nationalism*. London: Verso: 1981; Ernest Gellner. *Nations and Nationalism*. Ithaca, NY: Cornell University Press, 1983; Benedict Anderson. *Imagined Communities: Reflections on the Origin and Spread of Nationalism*. New York: Verso, 1991.
13. Étiene Balibar. "The Nation Form: History and Ideology." *New Left Review* XIII, no. 3 (1990), 334, quoted in Umut Özkırımlı. *Theories of Nationalism: A Critical Introduction*. New York: Palgrave, 2000, 199.
14. Rogers Brubaker. *Nationalism Reframed: Nationhood and the National Question in the New Europe*. Cambridge: Cambridge University Press, 1996, 15.
15. Leah Greenfeld. *Nationalism: Five Roads to Modernity* Cambridge, MA: Harvard University Press, 1992. Anthony Smith. *National Identity*. Reno: University of Nevada Press, 1991 distinguishes between Western civic-territorial and Eastern ethnic-genealogical nationalisms.
16. The exception that proves the rule in this case is Brubaker. *Nationalism Reframed*.
17. Sugar and Lederer, eds. *Nationalism in Eastern Europe* and Peter Sugar, ed. *Eastern European Nationalism in the 20th Century*. Washington DC: American University Press, 1995.
18. Jeremy King. "The Nationalization of East Central Europe: Ethnicism, Ethnicity, and Beyond," in Maria Bucur and Nancy M. Wingfield, eds. *Staging the Past: the Politics of Commemoration in Habsburg Central Europe, 1848 to the Present*. West Lafayette, IN: Purdue University Press, 2001.
19. Livezeanu, Irina. *Cultural Politics in Greater Romania: Regionalism, Nation Building, and Ethnic Struggle, 1918–1930*. Ithaca, NY: Cornell University Press, 1995. Timothy Snyder's *The Reconstruction of Nations: Poland, Ukraine, Lithuania, Belarus 1569–1999*. New Haven, CT: Yale University Press, 2003.
20. Keith Brown. *The Past in Question. Modern Macedonia and the Uncertainties of Nation*. Princeton, NJ: Princeton University Press, 2003; Katherine Verdery. *National Ideology under Socialism: Identity and Cultural Politics in Ceauşescu's*

Romania. Berkeley: University of California Press, 1991; Anastasia Karakasidou. *Fields of Wheat, Hills of Blood: Passages to Nationhood in Greek Macedonia, 1870–1990.* Chicago: University of Chicago Press, 1997; Pamela Ballinger. *History in Exile: Memory and Identity at the Borders of the Balkans.* Princeton, NJ: Princeton University Press, 2003.

21  Istvan Deak, J. Gross, and T. Judt, eds. *The Politics of Retribution in Europe: World War II and its Aftermath.* Princeton, NJ: Princeton University Press, 2000. Tony Judt. *Postwar: A History of Europe since 1945.* New York: Penguin, 2006.

22  Alexei Miller and Alfred J. Rieber, eds. *Imperial Rule.* Budapest: CEU Press, 2004. Carl L. Brown, ed. *Imperial Legacy: The Ottoman Imprint on the Balkans and the Middle East.* New York: Columbia University Press, 1996; Mark von Hagen and Karen Barkey, eds. *After Empire: Multiethnic Societies and Nation-building: The Soviet Union and Russian, Ottoman, and Habsburg Empires.* Boulder, CO: Westview Press, 1997. For a contrast, see Stefan Berger and Alexei Miller, eds. *Nationalizing Empires.* Budapest: CEU Press, 2015.

23  Greenfeld. *Nationalism,* 17, 25–26.

24  Ibid., 22–23.

25  See Maria Todorova. "The Course and Discourses of Bulgarian Nationalism," in Peter Sugar, ed. *Eastern European Nationalism in the Twentieth Century.* Washington DC: American University Press, 1995, who applies the methodology of Miroslav Hroch. *Social Preconditions of National Revival in Europe: A Comparative Analysis of the Social Composition of Patriotic Groups among the Smaller European Nations.* Cambridge: Cambridge University Press, 1985.

26  The Bulgarian term for the period of the national enlightenment is "Bulgarian Renaissance" (*Bulgasko Vuzrazhdane*).

27  Paisii Hilendarski and Petur Dinekov, ed. *Slavianobulgarska istoriia.* Sofia: Bulgarski Pisatel, 1972, 41–44.

28  Roger Gyllin. *The Genesis of the Modern Bulgarian Literary Language.* Stockholm: Uppsala, 1991, 13.

29  Elena Georgieva, S. Zherev, and V. Stankov, eds. *Istoriia na novobulgarskiia knizhoven ezik.* Sofia: Izdatelstvo na Bulgarskata Akademiia na Naukite, 1989, 45, 92–94.

30  Ibid., 93. *Diatopia* and *synopia* refer to the standardization of the dialectical basis of the language.

31  *Dunavski lebed* No. 7, 1860, cited in Georgieva. *Istoriia,* 173.

32  Liuben Karavelov is another of the great revolutionary figures who exerted a significant influence on the Bulgarian literary scene. The chairman of the Central Revolutionary Committee in Bucharest, he edited major émigré journals and newspapers like *Svoboda* and *Nezavisimost* and published novels and short stories.

33  Hroch. *Social Preconditions of National Revival in Europe,* 1985.

34 Gyllin. *The Genesis*, 29.
35 Ibid., 31, 108.
36 Ibid., 31.
37 Marin Pundeff. "The Bulgarian Academy of Sciences (On the Occasion of Its Centennial)." *East European Quarterly* III, no. 3 (1969), 372–73.
38 Aleksandur Velev. *Prosvetna i kulturna politika na pravitelstvoto na Aleksandur Stamboliiski.* Sofia: Izdatelstvo na BZNS, 1980, 96.
39 Autonomous Bulgaria was de facto independent after the Treaty of Berlin in 1878, but under Ottoman suzerainty. Formal independence was not proclaimed until 1908.
40 Liubomir Andreichin. *Iz istoriiata na nasheto ezikovo stroitelstvo.* Sofia: Narodna prosveta, 1977, 163–64.
41 See Mihail Arnaudov. *Istoriia na sofiiskiia universitet sv. Kliment Ohridski prez purvoto mu polustoletie: 1888–1938.* Sofia: Pridvorna pechatnitsa, 1939; Georgi Naumov, ed. *Istoriia na Sofiiskia universitet "Kliment Ohridski."* Sofia: Universitetsko Izdatelstvo "Kliment Ohridski," 1988.
42 Although the rector was technically a ministerial employee, the administrative council argued for his inviolability under the terms of the autonomous statute of the university.
43 Omarchevski had granted exceptions for editions of the works of Ivan Vazov, Georgi Rakovski, and a few others. For the blow-by-blow account that forms the basis of the narrative, see Naumov. *Istoriia*, 104–12. Arnaudov's *Istoriia* is indispensable as a collection of primary sources because it includes the relevant correspondence in full.
44 Bell. *Peasants in Power*, 180.
45 Naumov. *Istoriia*, 107.
46 Ibid., 108–09.
47 Velev. *Prosvetna i kulturna politika na pravitelstvoto na Aleksandur Stamboliiski*, 100–03.
48 Rusin Rusinov. *Istoriia na bulgarskiia pravopis.* Sofia: Nauka i Izkustvo, 1985, 86.
49 Velev. *Prosvetna i kulturna politika na pravitelstvoto na Aleksandur Stamboliiski*, 99.
50 Rusinov. *Istoriia na bulgarskiia pravopis*, 90.
51 Velev. *Prosvetna i kulturna politika na pravitelstvoto na Aleksandur Stamboliiski*, 99–100.
52 Velev's description is of a later debate, March 1922, when the same ministerial council affirmed its previous decision concerning the Ѫ. The occasion was the transformation of the Orthographic Directive into Law.
53 Ibid., 102.
54 Rusinov. *Istoriia na bulgarskiia pravopis*, 93.
55 Andreichin. *Iz istoriiata na nasheto ezikovo stroitelstvo*, 167.

56 Rusinov. *Istoriia na bulgarskiia pravopis*, 95.
57 Ibid., 93.
58 Petrova. *Bulgarskiiat Zemedelski Naroden Suiuz*, 95.
59 Rusinov. *Istoriia na bulgarskiia pravopis*, 92–93; Velev. *Prosvetna i kulturna politika na pravitelstvoto na Aleksandur Stamboliiski*, 103, 125–28.
60 Velev. *Prosvetna i kulturna politika na pravitelstvoto na Aleksandur Stamboliiski*, 103–04, 109, 111, 122.
61 Rusin Rusinov. *Istoriia na novobulgarskiia knizhoven ezik*. Veliko Turnovo: Pechat "Abagar," 1999, 444. An example is the article by Georgi Bakalov "Is the orthographic question moot?" (1927) that examines the Tsankov orthography as a means to isolate the workers from political life and cultural benefits. Rusinov. *Istoriia na bulgarskiia pravopis*, 107.
62 The early phase of Fatherland Front of 1945 was a coalition of Left of Center parties that along with the communists included the BANU.
63 Partha Chatterjee. "Whose Imagined Community?" *Millenium: Journal of International Studies* 20, no. 3 (Winter 1991), 521–26. Partha Chatterjee. *The Nation and Its Fragments*. Princeton, NJ: Princeton University Press, 1993.
64 Dipesh Chakrabarty. *Provincializing Europe*. Princeton, NJ: Princeton University Press, 2007.
65 Sumit Sarkar. "The Decline of the Subaltern in *Subaltern Studies*," in Sumit Sarkar, *Writing Social History*. Delhi: Oxford University Press, 1997, 82–108 and Vivek Chibber. *Postcolonial Theory and the Specter of Capital*. London: Verso, 2013.
66 A video of the debate can be found at http://www.youtube.com/watch?v=xbM8HJrxSJ4.
67 Chatterjee. *The Nation and Its Fragments*, 5.
68 Ibid., 5–6.
69 Book review by Daniel Clayton. *Progress in Human Geography* 28, no. 2 (April 2004), 267–68. Dipesh Chakrabarty. *Habitations of Modernity: Essays in the Wake of Subaltern Studies*. Chicago, IL: University of Chicago Press, 2002.
70 Katherine Verdery and Sharad Chari. "Thinking between the Posts: Postcolonialism, Postsocialism, and Ethnography after the Cold War." *Comparative Studies in Society and History* 51, no. 1 (2009), 6–34.
71 Maria Todorova. "Balkanism and Postcolonialism, or On the Beauty of the Airplane View," in Costica Bradatan and Sergei Oushakine, eds. *In Marx's Shadow: Knowledge, Power, and Intellectuals in Eastern Europe and Russia*. Lanham, MD: Lexington Books, 2010.
72 This move mimics the semiotic expansion and recapture of "postcolonial."
73 Verdery and Chari. "Thinking between the Posts," 11.
74 Chatterjee. *The Nation and Its Fragments*, 4.

75  Partha Chatterjee. "Subaltern History," in Neil Smelser and Paul Bates, eds. *International Encyclopedia of the Social and Behavioral Sciences*. Amsterdam: Elsevier Science, 2001, 15237–41.
76  A video of the debate can be found at http://www.youtube.com/watch?v=xbM8HJrxSJ4.
77  Chatterjee. *The Nation and Its Fragments*, 158.
78  Ibid.
79  Rather than marking a territory between the West and East (Russia and Germany) as per Timothy Snyder's problematic term "bloodlands," I would propose that a much larger and more stimulating revision of the geography of Europe could be achieved through marking the radical potential of the countries of Europe with a peasant majority as interwar agrarian Europe. See Timothy Snyder. *Bloodlands: Europe between Hitler and Stalin*. New York: Basic Books, 2010.
80  Lynne Viola, V. P. Danailov, N. A. Ivnitskii, and Denis Kozlov. *The War against the Peasantry 1927–1930: The Tragedy of the Soviet Countryside*. New Haven, CT: Yale University Press, 2005.

# Chapter 4

1  Original in НБКМ БИА, ф. 405, а.е. 2, л. 3, reprinted in Biliarski, ed. *BZNS, Aleksandur Stamboliiski i VMRO*, 377. See also ЦДА (Централен Държавен Архив), Fond 361k, opis 1, a.e. 12, f.149 back.
2  TsDA, Fond 361k, opis 1, a.e. 1, *istoricheska beleshka*.
3  Josef Kolář. *Bulharská demokratická emigrace v Československu v letech 1923-1933*. Prague: Academia, 1983, 49.
4  Kostadin Gurdev. "Bulgarskata zemedelska politicheska emigratsiia v Chehoslovakia prez 20-te i 30-te godini na XX vek," in Milen Kumanov and Rumiana Katsarova, eds. *Zemedelskoto dvizhenie v Bulgaria: istoriia, razvitie, lichnosti*. Pazardzhik: Belloprint, 2004, 214.
5  Ibid., 216.
6  *zdruzheni* (literally unified) has a particular agrarian connotation that derives from the local organizational unit of the BANU, the *druzhba*.
7  ЦДА (Централен Държавен Архив), Fond 361k, opis 1, a.e. 52, f. 3.
8  Ibid., 5.
9  ЦДА (Централен Държавен Архив), Fond 361k, opis 1, a.e. 52, f. 18–19 back. Although addressed to the temporary committee, this letter was read and annotated by the Representation Abroad. It even added a note to it stating that it had been answered via the June 20 general reply of Stancho Trifonov.
10  Ibid., 21.

11  ЦДА (Централен Държавен Архив), Fond 361k, opis 1, a.e. 53, f. 67–68.
12  *Zemledelets (L'Agrarien)*. Year 2, Issue 1–2, 1929, in ЦДА (Централен Държавен Архив), Fond 361k, opis 1, a.e. 74, f. 42 back.
13  *Bulletin MAB (3)* 1925, 3.
14  *Bulletin MAB (1)* 1926, 8–12.
15  *Bulletin MAB (2)* 1928, 100.
16  Charles Tilly. "Did the Cake of Custom Break," in John Merriman, ed. *Consciousness and Class Experience in Nineteenth Century Europe*. New York: Holmes & Meier, 1979, 34.
17  Henri Mendras and Amy Jacobs. "The Invention of the Peasantry: A Moment in the History of Post-World War II French Sociology." *Revue française de sociologie* 43, no. 1 (Année 2002), 157–71.
18  G. M. Dimitrov. "Agrarianism," in Feliks Gross, ed. *European Ideologies: A Survey of 20th Century Political Ideas*. New York: Philosophical Library, 1948, 396–450.
19  Kubů, Lorenz, Müller, and Šouša, eds. *Agrarismus und Agrareliten in Ostmitteleuropa*, 17.
20  Francisco Panizza, ed. *Populism and the Mirror of Democracy*. New York: Verso, 2005.
21  Ionescu and Gellner. *Populism: Its Meaning and National Characteristics*.
22  Ibid., 114.
23  Ibid., 115.
24  Ibid., 167.
25  http://opus.kobv.de/euv/volltexte/2011/40/.
26  Johan Eellend. "Agrarianism and Modernization in Inter-War Eastern Europe," in Piotr Wawrzeniuk, ed. *Societal Change and Ideological Formation among the Rural Population of the Baltic area 1880–1939*. Huddinge: Södertörns högskola, 2008, 55.
27  *Politicheski partii ili suslovni organizatsii* was published first in 1909, and subsequently underwent numerous editions. The following edition is used here: Aleksandur Stamboliiski. *Pulno subranie na uchineniiata mu*. Pod redaktsiata na Kuniu Kozhuharov, vol. III. Politicheski partii ili suslovni organizatsii, Sofia: Fondatsia Aleksandur Stamboliiski, 1947. For a sympathetic analysis of Stamnoliiski's views, see Bell. *Peasants in Power*, 59–73; Stefan Radulov. "Ideinite osnovi na zemedelskia rezhim," *Nauchni trudove na Akademiata za Obshtestveni Nauki i Sotsialno Upravlenie na TsK na BKP*. Seria istoriia, 105, Sofia, 1979, 7–76.
28  Stamboliiski. *Politicheski partii*, 176–209.
29  Ibid., 164–69.
30  Ibid., 112–32, 136–49.
31  Roumen Daskalov. *Ot Stambolov do Zhivkov. Golemite Sporove za Novata Bulgarska Istoriia*. Sofia: IK "Gutenberg," 2009, 131.

32  Ibid., 131. The BANU's legislation has been collected and published in: Aleksandur Stamboliiski. *Zakonodatelna deinost na pravitelstvoto na Bulgarskiia Zemedelski Naroden Suiuz, 1919-1923*. Sofia, 2003.
33  Bell. *Peasants in Power*, 183.
34  Raiko Daskalov. *Borba za Zemia*. Sofia, 1923, cited in Bell. *Peasants in Power*, 162–63.
35  On the land reform, see Bell. *Peasants in Power*, 164–68; Petrova. *Samostoiatelnoto upravlenie*, 114–23; Petrova. *Bulgarskiiat Zemedelski Naroden Suiuz*, 84–86; Richard Crampton. *Aleksandur Stambiliiski. Bulgaria*. London: House Histories, 2009, 114.
36  Bell. *Peasants in Power*, 166.
37  Ivan Ganev. *Agrarnata reforma v chuzhbina i u nas*. Sofia: Pechatnitsa "G. Provadaliev," 1946.
38  Bell. *Peasants in Power*, 171–76; Petrova. *Samostoiatelnoto upravlenie*, 103–08; Petrova. *Bulgarskiiat Zemedelski Naroden Suiuz*, 81–83, gives slightly different figures.
39  The female labor service, however, was implemented only to a limited extent and was repealed after the coup of June 9, 1923 (Bell. *Peasants in Power*, 175–76).
40  Petrova. *Samostoiatelnoto upravlenie*, 107–08.
41  Cited in Petrova. *Bulgarskiiat Zemedelski Naroden Suiuz*, 82.
42  Frederick B. Chary. *The History of Bulgaria*. Santa Barbara, CA: Greenwood, 2011, 60. See also Max Lazard. *Compulsory Labor Service in Bulgaria*. Geneva: International Labour Office, 1922. Professor Max Lazar, a French expert, was sent by the League of Nations to inspect the operations of the CLS and he judged positively its social benefits.
43  Sava Botev. Iosif Kovachev. *Zemledelieto v Bulgaria*. Sofia: Pechatnitsa S. M. Staikov, 1934, 607–40. See also Petko Dichev. *Za bulgarskoto selsko stopanstvo i za kooperatsiata v 40 godini. Sbornik ot trudove 1890-1930*. Sofia: Kooperativna pechatnitsa "Gutenberg," 1933.
44  *Zemledelski koperator*, 1, N.1–2, 7–8 (December 10, 1921, January 7, 1922), cited in Bell, *Peasants in Power*, 169.
45  Petrova. *Bulgarskiiat Zemedelski Naroden Suiuz*, 86.
46  Bell. *Peasants in Power*, 170.
47  Mary C. Neuburger. *Balkan Smoke: Tobacco and the Making of Modern Bulgaria*. Ithaca, NY: Cornell University Press, 2013, 117.
48  For the persistence of the cooperative model, see the data in *Statisticheski Godishnik na Tsartsvo Bulgaria 1942*, XXXIV. Sofia: Glavna Direktsia na Statistikata, 1942, 628–39. It was primarily this tradition that prompted Rumen Avramov to coin the framework of "communal capitalism" as a hindrance to the "properly functioning capitalism" of the West in his magnum opus *Komunalniat kapitalizum*. T.1–3. Sofia: Bulgarska nauka i kultura, Tsentur za liberalni strategii, 2007.

49 Petrova. *Bulgarskiiat Zemedelski Naroden Suiuz*, 126–29; Petrova. *Bulgarskiiat Zemedelski Naroden Suiuz*, 87–88; Bell. *Peasants in Power*, 171.
50 Editorial in *Zemedelsko zname*, XVIII, N. 43, March 6, 1922, 1.
51 Bell. *Peasants in Power*, is a strong proponent of the modernizing activities of the agrarians. Richard Crampton, on the other hand, characterizes Stamboliiski as a "reluctant modernizer" in "Modernization: Conscious, Unconscious and Irrational," in Ronald Schönfeld, ed. *Indistrialisierung und gesellschaftlicher Wandel in Südosteuropa*. München: Südosteuropa Gesellschaft, 1989, 128–29. Others deny altogether the modernizing potential of what they see as a regime bent on preserving small landowning. For the whole gamut of interpretations on Stamboliiski's reforms, see Daskalov. *Ot Stambolov do Zhivkov*, 136-61, here especially 157–58.
52 Crampton. *Aleksandur Stamboliiski*, 118–19.
53 Petrova. *Samostoiatelnoto upravlenie*, 109.
54 Bell. *Peasants in Power*, 168.
55 Ibid., 177; Petrova. *Samostoiatelnoto upravlenie*, 131–32. Bell and Petrova differ on the final numbers.
56 Bell. *Peasants in Power*, 178.
57 Svetla Baloutzova. *Demography and the Nation: Social Legislation and Population Policy in Bulgaria, 1918-1944*. Budapest: Central European University Press, 2011, 44.
58 Ibid., 47.
59 Ibid., 53.
60 Jackson, Jr. *Comintern and Peasant in East Europe*.
61 Goranovich. *Krah zelenogo internatsionala*. For Goranovich the organization is a reactionary formation that cleared the way for fascism.
62 Miller. *Forging Political Compromise*, 95.
63 *Bulletin Mezinárodního Agrárního Bureau*. 1923 (October–November), 3–8.
64 Ibid., 4.
65 Ibid., 6.
66 Ibid., 7.
67 *Bulletin Mezinárodního Agrárního Bureau*. 1925, issue 1, 16.
68 Ibid., issue I (V–VI), 56.
69 Ibid., issue I (V–VI), 56–57.
70 *Politicheski partii ili suslovni organizatsii* reprinted in, Kunio Kozhuharov, ed. *Aleksandur Stamboliiski: suchineniia (Tom 3 Trudova Demokratsiia)*. Sofia: Fondatsiia "Al. Stamboliiski," 1947.
71 *Bulletin Mezinárodního Agrárního Bureau*. 1925, issue I (V–VI), 58.
72 Ibid., issue II, 20.
73 For the *Front Paysan* in the 1930 see Robert Paxton. *French Peasant Fascism: Henry Dorgères' Greenshirts and the Crises of French Agriculture, 1929-1939*. New York: Oxford University Press, 1997.

74 *Bulletin Mezinárodního Agrárního Bureau.* 1925, issue III, 9.
75 Ibid., 21.
76 Both the Czech and the French texts leave unclear what exactly the "which" qualifies. From the context, it is more likely to refer to agrarianism and pacifism, rather than an anti-Bolshevik policy. *Bulletin Mezinárodního Agrárního Bureau.* 1925, issue III, 24.

# Chapter 5

1 Biondich. *Stjepan Radić*, 200.
2 *Priaporets* June 11, 1932, issue 128.
3 *Priaporets* June 6, 1923, Br. 125.
4 An example is the article "The Meaning of the Bulgarian 'Coup,'" in *Radikal.* June 15, 1923. Br. 131. "This is how the Belgrade newspaper *Vreme* titles for the third day in a row its headlining article about our 'coup.' *Vreme* was one of the staunchest supporters of Stamboliiski and that is why it still cannot come to terms with the fact that the Bulgarian dictator left the stage [sic]. The Belgrade newspaper searches for 'arguments' everywhere to claim that the change (or as it says, the 'coup') in Bulgaria is the work of those supporting the Macedonian cause [македонствующите] and that Bulgarian politics is returning towards 1915."
5 *Narod* June 12, 1923, issue 128.
6 *Epoha* June 13, 1923, issue 262.
7 *Radikal* June 11, 1923, issue 127. *Narod* June 11, 1923, issue 127.
8 *Otechestvo* June 15, 1923, issue 113. *Slovo* June 13, 1923, issue 321. *Narod* June 13, 1923, issue 129. *Radikal* June 13, 1923, issue 129.
9 The figure of 120 million leva was generally accepted at the time, but it does not correspond to the exchange rate of the *Wall Street Journal* from March 15, 1923 which valued 1 Swiss franc at 25.20 leva and which means that the sum was closer to 100 million.
10 *Radikal* June 13, 1923, issue 129. To provide more context, the next official announcement that permeated the press was an elaborate lie that described the shooting of Stamboliiski during an escape attempt.
11 Ibid.
12 *Otechestvo* June 15, 1923, issue 113. A poster containing a photograph of the money can be seen in ЦДА (Централен Държавен Архив), Fond 1335K, opis 1, a.e. 78, f. 2. Mihail Genovski in his *Aleksandur Stamboliiski ot blizo i otdaleko.* Sofia: Izdatelstvo na BZNS, 1982, 323, challenges the veracity of the photograph. He argues that the "Turlakov treasury bonds" that can be seen there had been out of circulation already before the coup and that therefore this picture is a montage.

13  A literal translation of *Druzhbashkiia Rezhim* would be "the regime of the united agriculturalists." The name is a derivative of the organizational units of the BANU, the *druzhbi*. Even though the agrarians would refer to themselves as *sdruzheni zemedeltsi* (united agriculturalists), *druzhbashi* was a derogatory appellation. In its referral to people based on their organizational unit, the term is close to the (mis) use of the term *soviet*.

14  *Druzhbashkiia Rezhim—Dokumenti: Chuzhdiia Pechat za ministerskata promiana na 9 Iunii i za upravlenieto na Bulgarskiia Zemedelski Suiuz, Book 7*. Sofia: Izdanie na Komisiiata za Preglezhdane Arhivite na Bivshite Ministri, 1923. *Druzhbashkiia Rezhim—Dokumenti: Chuzhdiia Pechat za Druzhbashtinata, Book 10*. Sofia: Izdanie na Komisiiata za Preglezhdane Arhivite na Bivshite Ministri, 1923.

15  *Druzhbashkiia Rezhim—Dokumenti Partizanskata gangrena v Uchilishteto, Book 3*. Sofia: Izdanie na Komisiiata za Preglezhdane Arhivite na Bivshite Ministri, 1923.

16  *Druzhbashkiia Rezhim—Dokumenti: Aleksandur Stamboliiski, Books 1 and 2*. Sofia: Izdanie na Komisiiata za Preglezhdane Arhivite na Bivshite Ministri, 1923.

17  *Druzhbashkiia Rezhim—Dokumenti: Aleksandur Stamboliiski, Book 2*. Sofia: Izdanie na Komisiiata za Preglezhdane Arhivite na Bivshite Ministri, 1923, 46–47.

18  Ibid., 39–44.

19  Ibid., 31.

20  Ibid., 33.

21  ЦДА (Централен Държавен Архив), Fond 285, opis 1, a.e. 318, f. 226.

22  *Bulgarski Periodichen Pechat 1844–1944*. Sofia: Nauka i Izkustvo, 1969.

23  *Zemledelska Zashtita* March 30, 1924, issue 38.

24  *Zemledelsko Zname (Prague)* December 14, 1923, issue 10.

25  Ibid., 3.

26  Ibid., 4. *Zemledelsko Zname* identifies the author of the *Druzhbashkia Rrezhim* brochures on Stamboliiski as Krustiu Stanchev, the editor of *Nezavisimost*, the organ of the National-Liberal Party.

27  *Zemledelsko Zname (Prague)*, June 14, 1924, issue 30.

28  Ibid., 4. Tsar Boris did visit Stamboliiski at his house in Slavovitsa on June 7, 1923, supposedly to convey his best wishes, even though he was informed of the coup. Petrova. *Aleksandur Stamboliiski*, 45.

29  *Zemledelsko Zname (Prague)*, December 15, 1924, issue 8–9.

30  A. Grebenarov. "Pokazaniia na gen. Ivan Vulkov i kap. Ivan Harlakov za subitiiata okolo 9 iuni 1923 g." *Makedonski Pregled*. Book 2, 2003, 107–16.

31  ЦДА (Централен Държавен Архив), Fond 768К, оп. 1, a.e. 109, f. 1–3—Razpit na I. Harlakov za ubiistvoto na Al. Stamboliiski pred narodniia sud. Sofia, February 8, 1945.

32  Ibid., 3. Mihail Genovski confirms this in *Aleksandur Stamboliiski ot blizo i otdaleko*, 324, where he writes that the money was destined to members of the Reparations Committee in return for a diminution and deferral of payments.

33 The eighteen million most likely refers to the value in leva of all the banknotes found.
34 АМВР (Архив на Министерството на Вътрешните Работи), II сл. 406, т. XXVI, 153–71 (back)—Protokol za razpit, 7 Sept. 1946. This is much more detailed statement than the one Grebenarov could consult in Централен военен архив—В. Търново, Fond 23, opis I, a. e. 823, f. 81–83.
35 Alexander Grebenarov. "VMRO i SSSR—Sblizhenie s predizvesten krai (1923–1924)," in *Mezhdunarodna Nauchna Konferentsiia 'Bulgariia I Rusiia mezhdu priznatelnostta i pragmatizma—Dokladi*. Sofia, March 2008. This information is based on the memoirs of Stoicho Moshanov and Georgi Bazhdarov, the external representative of IMRO.
36 АМВР (Архив на Министерството на Вътрешните Работи), II сл. 406, т. XXVI, 164.
37 ЦДА (Централен Държавен Архив), Fond 285, opis 1, a.e. 318, f. 226.
38 ЦДА (Централен Държавен Архив), Fond 285, opis 1, a.e. 319, f. 87 (back).
39 ЦДА (Централен Държавен Архив), Fond 285, opis 3, a.e. 23—April 4, 1924.
40 Petrova. *Samostoiatelnoto upravlenie*, 384–86.
41 ЦДА (Централен Държавен Архив), Fond 1335K, opis 1, a.e. 77, f. 1.
42 Ibid.
43 The decision and argumentation of the court, as well as the response to an appeal, can be found in ЦДА (Централен Държавен Архив), Fond 1335K, opis 1, a.e. 77, f. 1–47.
44 Ibid., 19.
45 Ibid., 20. Note: The court documents constantly employ the phrase "обявени за заподозрени в незаконно забогатяване" (declared to be suspected of illegal enrichment) to refer to Stamboliiski and Boiadzhiev. In this sentence, however, "заподозрени" (suspected) is suspiciously absent, which further emphasized the complete disregard for the presumption of innocence. I have corrected this convenient omission, in order to better emphasize the travesty of the court's reasoning.
46 Ibid., 24–26, 43–44. The court's acceptance of 3,594,029 leva was an improper calculation and after an appeal forced it to reexamine its calculations, and the sum was revised to 4,689,000.
47 Ibid., 25. Mihail Genovski. *Aleksandur Stamboliiski ot blizo i otdaleko*, 325, describes the villa as being a two-story structure with a wine cellar on the bottom and two rooms and a veranda on the top.
48 Ibid., 26.
49 Ibid., 43–47.
50 Petrova. *Samostoiaelnoto upravlenie*, 28.
51 For further information see: Bell. *Peasants in Power*, 1977. Барев, Ценко. *Принос към Историята на Българския Земеделски Народен Съюз: Борба, Идеология, Принципи*. София: Булвест 2000, 1994.

52 Petrova. *Samostoiaelnoto upravlenie*, 17–23.
53 See for example, Nathaniel Leff. "Economic Development through Bureaucratic Corruption." *American Behavioral Scientist* 8, no. 3 (November 1964), 8–14; James C. Scott. *Comparative Political Corruption*. Englewood Cliffs, NJ: Prentice-Hall, 1972.
54 Susan Rose-Ackerman. *Corruption: A Study in Political Economy*. New York: Academic Press, 1978.
55 Mény Yves. "'Fin de siècle' Corruption: Change, Crisis and Shifting Values." *International Social Science Journal* 48, no. 3 (1996), 309–20.
56 "Transparency International's *Corruption Perceptions Index (CPI)* is the best known of our tools. First launched in 1995, it has been widely credited with putting the issue of corruption on the international policy agenda. The CPI ranks almost 200 countries by their perceived levels of corruption, as determined by expert assessments and opinion surveys." http://www.transparency.org/policy_research/surveys_indices/cpi.
57 *Rapport Mondial sur la Corruption 2004: Thème Special:la corruption politique*. Paris: Éditions Karthala, 2004. See also Lucas Achathaler, Domenica Hofmann, and Mattias Pazmandy, eds. *Korruptionsbekämpfung als globale Herausforderung: Beiträge aus Praxis und Wissenschaft*. Wiesbaden: VS Verlag, 2011.
58 Arvind K. Jain. *The Political Economy of Corruption*. New York: Routledge, 2001. See also Donatella Della Porta and Alberto Vannucci. *Corrupt Exchanges: Actors, Resources, and Mechanisms of Political Corruption*. New York: Aldine de Gruyter, 1999 and Alan Doig and Robin Theobald, eds. *Corruption and Democratization*. Portland, OR: Frank Cass, 2000.
59 Doig and Theobald, eds. *Corruption and Democratization*.
60 Ibid., 1.
61 Ibid., 6–8.
62 Ibid., 149.
63 Leslie Holmes. *The End of Communist Power: Anti-corruption Campaigns and Legitimation Crisis*. New York: Oxford University Press, 1993. Leslie Holmes. *Rotten States? Corruption, Post-Communism, and Neoliberalism*. Durham, NC: Duke University Press, 2006.
64 Betty Glad and Shiraev Eric, eds. *The Russian Transformation: Political, Sociological, and Psychological Aspects*. New York: St. Martin's Press, 1999. Daniel Smilov and Jurij Toplak. *Political Finance and Corruption in Eastern Europe: The Transition Period*. Aldershot: Ashgate, 2007. See also Rashma Karklins, *The System Made Me Do It: Corruption in Post-Communist Societies*. Armonk: M. E. Sharpe, 2005.
65 Smilov and Toplak. *Political Finance and Corruption in Eastern Europe*, 1.
66 Ibid.
67 Glad and Shiraev, eds. *The Russian Transformation: Political, Sociological, and Psychological Aspects*, 119. The reference is to Scott. *Comparative Political Corruption*.

68 Ibid., 128.
69 Harry G. West and Parvathi Raman. *Enduring Socialism: Explorations of Revolution & Transformation, Restoration & Continuation*. New York: Berghahn Books, 2009, 3. The broader context of the quotation is as follows: "Implicit in these studies was the idea that progression to democracy required a decisive break with the practices of the socialist state. The sudden collapse of the socialist world was thus commonly viewed through an entropic lens, where the measure of internal disorder and decay was not always outwardly visible, but the system was nevertheless rotting from within. By extension, this body of work also laid the foundation for fieldwork that sought to chart the successes and failures of transition, asking what had gone wrong, and how it could be remedied."
70 Dieter Haller and Cris Shore. *Corruption: Anthropological Perspectives*. Ann Arbor: Pluto Press, 2005.
71 Ibid., 9–10.
72 Italo Prado. *Between Morality and the Law: Corruption, Anthropology and Comparative Society*. Aldershot: Ashgate, 2004, 15. See also West and Raman. *Enduring Socialism*, 2009.
73 Ruth A. Miller. *The Erotics of Corruption: Law, Scandal, and Political Perversion*. Albany, NY: State University of New York Press, 2008, xv.
74 Tina Rosenberg. *The Haunted Land: Facing Europe's Ghosts after Communism*. New York: Vintage Books, 1996, 126.
75 "Obituary: Ramiz Alia (1925–2011), Albania's Last Communist Leader." http://suite101.com/article/obituary-ramiz-alia-1925-2011-albanias-last-communist-leader-a392229#ixzz26OKPp3Jp.
76 http://www.nytimes.com/1992/07/30/world/honecker-flown-to-berlin-to-face-criminal-trial.html?pagewanted=all&src=pm and A. James McAdams. "The Honecker Trial: The East German past and the German Future." The Helen Kellogg Institute for International Studies, Working paper #216—January 1996. http://kellogg.nd.edu/publications/workingpapers/WPS/216.pdf.
77 The following account is from Vladko Maček. *Memoari*. Zagreb: Dom i Svijet, 2003, 99–100.
78 Davidović formed a cabinet on July 24, 1924, after Pašić tendered his resignation on April 12, 1924.
79 Maček. *Memoari*, 100.
80 Wayne Vucinich and Jozo Tomasevich. *Contemporary Yugoslavia: Twenty Years of Socialist Experiment*. Berkeley: University of California Press, 1969, 16.
81 Maček. *Memoari*, 102–03.
82 Quoted in Biondich. *Stjepan Radić*, 193. The quote itself is from Krizman, ed. *Korespondencija Stjepana Radića 1885–1928*, vol. 2, 579.
83 Reviakina. *Kominternut*, 91.

84 Ibid., 92.
85 Biondich. *Stjepan Radić*, 199.
86 Ibid.
87 Quoted in Luiza Reviakina, Luchezar Stoianov, and Volodya Milachkov. *Kominternut i Bulgariia (Mart 1919-septemvri 1944 g.) Dokumenti*. Sofia: Glavno upravlenie na Arhivite, 2005, 267. Referring to a document in *Российского государственного архива социально-политической истории* (РГАСПИ, ф. 535, оп. 1, д. 21, л. 9).
88 Reviakina. *Kominternut*, 87.
89 Arhiv Jugoslavije. Seljačka Internacionala—Krestintern Fond 519, 1924. Godina. Arhivna Jedinica 49.
90 Arhiv Jugoslavije. Seljačka Internacionala—Krestintern Fond 519, 1924. Godina. Arhivna Jedinica 43.
91 Reviakina. *Kominternut*, 103.
92 Maček. *Memoari*, 105.
93 Ivan Mužić. *Stjepan Radić u Kraljevini Srba, Hrvata i Slovenaca*. Ljubljana, Zagreb: Hrvatsko književno društvo sv. Ćirila i Metoda, 1987, 199.
94 Richard Crampton. *Eastern Europe in the Twentieth Century—and After*. 2nd ed. New York: Routledge, 1997, 136. Joseph Rothschild. *East Central Europe between the Two World War*. Seattle: University of Washington Press, 1974.
95 Matković. *Povijest Hrvatske Seljačke Stranke*, 201. See also Bosiljka Janjatović. *Stjepan Radić: Progoni—Zatvori—Suđenja—Ubojstvo, 1889-1928*. Zagreb: Dom i Svijet, 2003.
96 The bitterness against Radić's "betrayal" was still visible in 1928 in a resolution of the CC of the Bulgarian Communist Party. "Also the agrarian leadership on the right clearly takes positions against the Communist Party, and how will they act in the case that they enter into a future government is best shown by the example of Radić: despite his leftist fanfaronades and revolutionary overtures, once come to power, he no longer thought about the law for the defense of the nation." Reviakina. *Kominternut i Bulgariia*, 569; ЦДА (Централен Държавен Архив), Fond 3 Б. opis. 4, a.e. 187, f. 77–82.

# Chapter 6

1 Crampton. *Eastern Europe in the Twentieth Century—and After*, 1997.
2 Barbara Jelavich. *History of the Balkans: Twentieth Century*, vol. 2. Cambridge: Cambridge University Press, 1983, 185.
3 Ivan Berend. *Decades of Crisis: Central and Eastern Europe Before World War II*. Berkeley: University of California Press, 1998.

4 Gail Kligman and Katherine Verdery. *Peasants under Siege: The Collectivization of Romanian Agriculture, 1949-1962*. Princeton, NJ: Princeton University Press, 2011; Gerald Creed. *Domesticating Revolution: From Socialist Reform to Ambivalent Transition in a Bulgarian Village*. University Park: Pennsylvania State University Press, 1998; Melissa Bokovoy. *Peasants and Communists: Politics and Ideology in the Yugoslav Countryside, 1941-1953*. Pittsburgh: University of Pittsburgh Press, 1998.
5 Kligman and Verdery. *Peasants under Siege*, 6.
6 Aleksandur Stamboliiski. *Printsipite na bulgarskiiat zemledelski suiuz (pisano v zatvora)*. Sofia: Pechatnitsa na Z. D. Vidolov, 1919, 8.
7 Ibid., 8-9.
8 Evgenia Kalinova and Iskra Baeva. *Bulgarskite prehodi 1939-2005*. Sofia: Paradigma, 2005, 51.
9 *Politika*, N.6, October 17, 1944, cited in: Rumiana Bogdanova. "Zemedelskoto dvizhenie sled vtorata svetovna voina (Putiat na partiinoto obezlichavane)," in Georgi Markov, ed. *Zemedelskoto dvizhenie v Bulgaria: Istoriia, razvitie, lichnosti*. Pazardhik: Regionalen istoricheski muzei, Institut po istoriia pri BAN, 2004, 264.
10 Kalinova and Baeva. *Bulgarskite prehodi*, 60-61.
11 Charles Moser. *Dimitrov of Bulgaria. A Political Biography of Dr. G. M. Dimtrov*. Thornwood, NY: Caroline House Publishers, 1979; Dr. G. M. Dimitrov. *Spomeni*. Sofia, 1993.
12 Michael M. Boll. *Cold War in the Balkans: American Foreign Policy and the Emergence of Communist Bulgaria 1934-1947*. Lexington: University of Kentucky Press, 1984, 186.
13 Zhivko Popov. *Ubitite zaradi ideite si. Politicheskiat vuzhod i zhiteiskoto krushenie na familia Petkovi*. Sofia: Kira 21, 2004; Georgi Gunev. *Kum brega na svobodata ili za Nikola Petkov i negovoto vreme*. Sofia: Infornatsionno obsluzhvane AD, 1989, 1992; Kalin Iosifov. *Totalitarnoto nasilie v bulgarskoto selo (1944-1951) i posleditisite za Bulgaria*. Sofia: Universitetsko Izdatelstvo "Kliment Ohridski," 1999, 2003; Veselin Stoianov, ed. *Predsmurtnite pisma na Nikola Petkov do Georgi Dimitorv i Vasil Kolarov*. Sofia: Universitetsko Izdatelstvo "Kliment Ohridski," 1992; Petur Semerdzhiev. *Sudebniat protses sreshtu Nikola Petkov*. Sofia: n.p., 1990.
14 Contrary to the oppositional tradition in the BANU to the IMRO, Moser's Party formed an electoral coalition in 2005 with IMRO-Bulgarian National Movement. That coalition, which at its height received 5.2 percent of the popular vote in the 2005 elections, is the Bulgarian National Union.
15 For a survey of post-1989 BZNS, see Iliana Marcheva, "Zemedelskoto dvizhenie i negovata sudba v politicheskata sistema na Bulgariia sled 1989 godina." in Kumanov and Katsarova, eds. *Zemedelskoto dvizhenie v Bulgariia*.
16 Prior, it was in the coalition, Democratic Left, with the Socialist Party in 1994 and 1997.

17  http://www.bznsas.org/images/doc/Ustav_ZSAS.pdf.
18  http://www.bznsas.org/images/doc/Programa_ZSAS.pdf.
19  Ibid.
20  Bokovoy. *Peasants and Communists*, 32–34; Sabrina P. Ramet. *The Three Yugoslavias: State-Building and Legitimation, 1918–2005*. Bloomington: Indiana University Press, 2006, 170–71.
21  Jozo Tomasevich. *War and Revolution in Yugoslavia, 1941–1945: Occupation and Collaboration*. Stanford, CA: Stanford University Press, 2001, 360–61.
22  Cited in Ramet. *The Three Yugoslavias*, 172.
23  *The Tablet* 193, no. 5678 (March 19, 1949), 180.
24  Ibid., 180.
25  Ibid., 181.
26  http://www.hss.hr/files/Osnovne%20odrednice%20programa%20HSS.pdf.
27  Ibid.
28  For coverage of the post-1989 political scene, consult Miroslav Mareš, Pavel Pšeja. "Agrarian and Peasant Parties in the Czech Republic: History, Presence and Central European Context." http://ispo.fss.muni.cz/uploads/2download/working_papers/Agrarian%20and%20Peasant%20Parties%20in%20the%20Czech%20Republic.pdf and Janusz Bugajski. *Political Parties of Eastern Europe: A Guide to Politics in the Post-Communist Era*. Armonk, NY: M. E. Sharpe, 2002.
29  For example, see Miroslav Pekník, ed. *Milan Hodža, politik a žurnalista*. Bratislava: VEDA Vydavateľstvo Slovenskej akadémie vied, 2008. Štefan Šebesta, ed. (Dissertation of Pavol Lukáč). *Milan Hodža v zápase o budúcnosť strednej Európy v rokoch 1939–1944*. Bratislava: VEDA Vydavateľstvo Slovenskej akadémie vied, 2005. Pekník, ed. *Milan Hodža: Statesman and Politician*.
30  Bradley F. Abrams. "The Second World War and the East European Revolution." *East European Politics and Societies* 16, no. 3 (Fall 2002), 623–64. Jan Gross's "War as Revolution" is the first chapter of Norman Naimark and Leonid Gabianskii, eds. *The Establishment of Communist Regimes in Eastern Europe, 1944–1949*. Boulder, CO: Westview Press, 1997. See also Melissa Bokovoy, "Peasants and Partisans: a Dubious Alliance" in the same volume.
31  The parties that remained outside were a wing of the National Socialists, the Social Democrats, and the Communists.
32  *Bulletin Mezinárodního Agrárního Bureau*, no. 4 (December 15, 1938), 130.
33  "In restructuring Czech and Slovak politics after the war Beneš was determined to eliminate the grip of the 'agrarian barons,' as he termed them, from the economy and politics. The Communist party was to serve as a counter-balance to any remaining conservative Republican interests. Beneš was to achieve his goals. The Košice Program of 1945, which set the agenda for the government of the newly re-created Czechoslovakia, prohibited the existence of the Republican Party, along

with certain other parties, such as the National Democrats, because of alleged complicity with the Germans." Miller. *Forging Political Compromise*, 203–04.

34 Miloš Parma, ed. *Agenti Zelené Internacionály: nepřátelé n aší vesnice*. Prague: Orbis, 1952, 16.
35 Ibid., 15.
36 See Petr Anev. "Procesy s údajnými přisluhovači Zelené internacionály." *Paměť a dějiny*. Roč 6, 2012, č. 4, 23–34.
37 ЦДА (Централен Държавен Архив), Fond 361K, opis 1, a.e. 87, f. 68 (back).
38 Ibid., f. 62.
39 Beran was sentenced together with Josef Tiso (death) on April 15, 1947.
40 ЦДА (Централен Държавен Архив), Fond 361K, opis 1, a.e. 87, f. 62 (back).

# Bibliography

## Archives

Belgrade, Arhiv Jugoslavije -
    Fond 14, Ministarstvo unutrašnjih poslova
    Fond 138, Ministarski savet
    Fond 519, Krestintern
    Fond 80, Jovan Jovanović Pižon

Prague, Arhív Národního muzea
    Pozůstalost Milana Hodži

Prague, Státní ústřední archív
    Fond 412, Republikánská strana
    Ministerstvo národní bezpečnosti
    Zemědělsko-lesnické oddělení
        Republikánská strana zemědělského a malorolnického lidu 1918-1939
        Svaz zemědělských průmyslníků a velkostatkářů

Sofia, ЦДА (Централен Държавен Архив), *Tsentralen Durzhaven Arhiv* (Central State Archives, TsDA)
    Fond 176K: Ministerstvo na vunshnite raboti i izpovedaniata
    Fond 264K: Ministerstvo na vutreshnite raboti i narodnoto zdrave
    Fond 361K: Zadgranichno predstavitelstvo na BZNS, 1922–1937
    Fond 370K: Direktsia na politsiata
    Fond 375K: Demokratichen sgovor
    Fond 276K: Bulgarski Gradinarski Suyuz
    Fond 304K: Bulgarska legatsia vuv Viena
    Fond 315K: Bulgarska legatsia v Belgrad
    Fond 382K: Bulgarska legatsia v Parizh
    Fond 383K: Bulgarska legatsia vuv Varshava
    Fond 391K: Bulgarska legatsia v Bratislava
    Fond 394K: Bulgarska legatsia v Zagreb
    Fond 393K: Bulgarska postoianna delegatsia pri Obshtestvoto na Narodite
    Fond 460K: Bulgarska legatsia v Praga

Fond 768K: Ivan Tuleshkov
Fond 1335K: Dimitur Peshev
Fond 255K: Aleksandur Stoimenov Stamboliiski, 1879–1923
Fond 561K: Nikola Petkov
Fond 792K: Aleksandur Obbov
Fond 494K: Petko D. Petkov
Fond 409K: Dimitur Gichev
Fond 422K: Vergil Dimov
Fond 1635: Yordan Pekarev
Fond 695K: Raiko Daskalov
Fond 698K: Dimitur Dragiev
Fond 152K: Grigor Cheshmedzhiev
Fond 1635K: Yurdan Pekarev
Fond 935K: Konstantin Muraviev
Fond 850K: Hristo Baev

Sofia, АМВР (Архив на Министерството на Вътрешните Работи)
Archives of the Ministry of Interior (AMVR)
    Fond 1: Tsentralno upravlenie
    Fond 2: Direktsia na politsiata
    Fond 6: Oblastna direkstia, Pleven
    Fond 9: Oblastna direkstia, Sofia
    Fond 14: Ministerstvo na pravosudieto i zatvorite
    Fond 15: Ministrestvo na prosvetata
    Fond 16: Durzhavna zhnadarmeria
    Fond 46: Sudebni instantsii
    Fond 68: Voenno-polevi sud, Sofia
    Fond 74: Oblasten voenen sud

Sofia, Архив на БАН (Archive of BAS) Arhiv na BAN
    Fond 73K: Petur Durvingov

Veliko Turnovo, Централен военен архив (Central Army Archive)

Zagreb, Arhiv Hrvatske
    Group VI, Građanske stranke
    Group XXI-P, Politička situacija
    Rukopisna Ostavština Antuna, Pavla i Stjepana Radića

Zagreb, Povilesni Arhiv
    Fond Stjepana Radića i HSS

Zagreb, Arhiv Instituta za Suvremenu Povijest
    Zbirka Gradjanskih Stranaka, Grupa VI-C, Hrvatska Seljačka Stranka

## Contemporary periodicals

Bulgaria:
*Epoha*
*Narod*
*Narodno Stopanstvo*
*Otechestvo*
*Politika*
*Priaporets*
*Radikal*
*Selski glas*
*Selski kooperator*
*Selski vestnik*
*Seyach*
*Slovo*
*Spisanie na zemedelskite kamari*
*Vreme*
*Zemedelsko-stopanski vuprosi*
*Zemledelska Zashtita*
*Zemledelski koperator*
*Zemledelsko Zname*

Czechoslovakia:
*Bulletin Mezinárodního Agrárního Bureau*

Yugoslavia:
*Demokratja*
*Hrvatska Smotra*
*Hrvatska Straža*
*Jedinstvo*
*Jugolsavenska Reć*
*Jugoslavenska straža*
*Jutarni List*
*Novosti*
*Politika*
*Samouprava*
*Selo*
*Slobodni Dom*

International:
*La fédération balkanique*
*Nineteenth Century*
*Pravda*

*The New York Times*
*The Tablet*

## Published documents and memoirs

Aleksandur Stamboliiski. *Zakonodatelna deinost na pravitelstvoto na Bulgarskiia Zemedelski Naroden Suiuz, 1919-1923.* Sofia, 2003.

*Annuaire international de legislation agricole.* (publie par l'Institut International d'Agriculture). Annees 1919 a 1924.

*Annuaire Internationale de Statistique Agricole 1930-1931.* Rom, 1931.

BKP. *Kominternut i makedonskiat vupros (1917-1946).* T.1. Sustaviteli Tsocho Biliarski i Iva Burilkova, *Arhivite Govoriat*, t.4. Sofia: Glavno upravlenie na arhivite pri ministrekia suvet, 1998.

BKP. *Kominternut i makedonskiat vupros (1917-1946).* T.2 Sust. Tsocho Biliarski i Iva Burilkova, *Arhivite Govoriat*, t.5. Sofia: Glavno upravlenie na arhivite pri ministrekia suvet, 1999.

BZNS. *Aleksandur Stamboliiski i VMRO: Nepoznatata voina.* Sustavitel Tsocho Bilairski. Sofia: Aniko, 2009.

*Convention pour la creation d'une societe international de credit hypothecaire agricole.* Geneva, 1931.

Dimitrov, Dr., G. M. *Spomeni.* Sofia, 1993.

*Druzhbashkiia Rezhim- Dokumenti: Aleksandur Stamboliiski, Book 1 and 2.* Sofia: Izdanie na Komisiiata za Preglezhdane Arhivite na Bivshite Ministri, 1923.

*Druzhbashkiia Rezhim—Dokumenti Partizanskata gangrena v Uchilishteto, Book 3.* Sofia: Izdanie na Komisiiata za Preglezhdane Arhivite na Bivshite Ministri, 1923.

*Druzhbashkiia Rezhim—Dokumenti: Chuzhdiia Pechat za ministerskata promiana na 9 Iunii i za upravlenieto na Bulgarskiia Zemedelski Suiuz, Book 7.* Sofia: Izdanie na Komisiiata za Preglezhdane Arhivite na Bivshite Ministri, 1923.

*Druzhbashkiia Rezhim—Dokumenti: Chuzhdiia Pechat za Druzhbashtinata, Book 10.* Sofia: Izdanie na Komisiiata za Preglezhdane Arhivite na Bivshite Ministri, 1923.

Grebenarov, A. "Pokazaniia na gen. Ivan Vulkov i kap. Ivan Harlakov za subitiiata okolo 9 iuni 1923 g." *Makedonski Pregled.* Book 2, 2003.

*Izvori za bulgarskata istoriia, t. XXIII: Chehoslovashki izvori za bulgarskata istoriia,* t.I, ed. Jozef Kolarzh, Ivan Shtovichek, Valerian Bistritski, Vasil Vasilev, Lidia Manolova. Sofia: Izdatelstvo na BAN, 1985.

*Izvori za bulgarskata istoriia, t. XXVII: Chehoslovashki izvori za bulgarskata istoriia,* t.II, ed. Jozef Kolarzh, Ivan Shtovichek, Valerian Bistritski, Vasil Vasilev, and Lidia Manolova. Sofia: Izdatelstvo na BAN, 1987.

*Izvori za bulgarskata istoriia, t. XXVIII: Chehski i slovashki izvori za bulgarskata istoriia,* t.III, ed. Jozef Kolarzh, Ivan Shtovichek, Valerian Bistritski, Vasil Vasilev, and Lidia Manolova. Sofia: Izdatelstvo na BAN, 1994.

*Izvori za bulgarskata istoriia, t. XXIX: Chehski i slovashki izvori za bulgarskata istoriia*, t.IV, ed. Jozef Kolarzh, Ivan Shtovichek, Valerian Bistritski, Veselin Starchevich, Miroslav Teihman, Vasil Vasilev, Krustiu Manchev, and Lidia Manolova. Sofia: Izdatelstvo na BAN, 2008.

Kavaldzhiev, Todor. *100 godini BZNS. Album spravochnik*. Sofia: Anubis, 2001.

*Kominternut i Bulgaria (Mart 1919-septemvri 1944 g.) Dokumenti*. T.1 Sust. Luiza Reviakina, Luchezar Stoianov, Volodia Milachkov, R. P. Grishina, and N. S. Lebedeva, *Arhivite Govoriat*, t.36. Sofia: Glavno upravlenie na arhivite pri ministerskia suvet, 2005.

*Kominternut i Bulgaria (Mart 1919-septemvri 1944 g.) Dokumenti*. T.2 Sust. Luiza Reviakina, Luchezar Stoianov, Volodia Milachkov, R. P. Grishina, and N. S. Lebedeva, *Arhivite Govoriat*, t.37. Sofia: Glavno upravlenie na arhivite pri ministerskia suvet, 2005.

Krizman, B. *Korespondencija Stjepana Radiča, 1919–1928*, t.I–II. Zagreb: Institut za hrvatsku povijest, 1973.

Maček, Vladko. *Memoari*. Zagreb: Dom i Svijet, 2003.

National Assembly. Stenographic journals (Народно Събрание. Стенографски дневници).

*Programi, programni dokumenti i ustavi na burzhoaznite partii v Bulgaria, 1879–1918*. Sofia: Nauka i izkustvo, 1992.

*Statisticheski Godishnik na Tsartsvo Bulgaria 1942*, XXXIV. Sofia: Glavna Direktsia na Statistikata, 1942.

Stefanova, Slava. *Mezhdunarodni aktove i dogovori (1648–1918)*. Sofia, 1958.

Stoianov, Veselin, ed. *Predsmurtnite pisma na Nikola Petkov do Georgi Dimitorv i VasilKolarov*. Sofia: Universitetsko Izdatelstvo "Kliment Ohridski," 1992.

## Secondary literature

*120 godini ot rozhdenieto na Aleksandur Stamboliiski*. Dokladi ot tematichnata konferentsia, organizirana ot BZNS – Naroden suiuz s pomoshtta na fondatsia "Hans Zidel" i fondatsia "XXI vek – Stara Zagora." Sofia: Parlamentarna grupa Naroden suiuz, 1999.

Abrams, Bradley F. "The Second World War and the East European Revolution." *East European Politics and Societies* 16/3 (Fall 2002), 623–64.

Achathaler, Lucas, Domenica Hofmann, and Mattias Pazmandy, eds. *Korruptionsbekämpfung als globale Herausforderung: Beiträge aus Praxis und Wissenschaft*. Wiesbaden: VS Verlag, 2011.

Agresti, Olivia Rossetti. *David Lubin: A Study in Practical Idealism*. Berkeley: University of California Press, 1941.

Althusser, Louis. "Ideology and Ideological State Apparatus (Notes Towards an Investigation)." In Louis Althusser, ed. *Lenin and Philosophy and Other Essays*. New York: Monthly Review Press, 2001, 85–132.

Amin, S., and K. Vergopoulos. *La question paysanne et le capitalisme.* Paris: Éditions Anthropos-Idep, 1974.

Anderson, Benedict. *Imagined Communities: Reflections on the Origin and Spread of Nationalism.* New York: Verso, 1991.

Andreichin, Liubomir. *Iz istoriiata na nasheto ezikovo stroitelstvo.* Sofia: Narodna prosveta, 1977.

Anev, Petr. "Procesy s údajnými přisluhovači Zelené internacionály." *Paměť a dějiny* 6/4 (2012), 23–34.

Angonov, K. V., ed. *Evropeiskii soiuz, agrarnie konferentsii i regionalnie soglashenia.* Moskva, 1932.

Anina, D. "Sotsial'noe polozhenie krest'ianstva v Yugoslavii." *Krestianskii internatsional,* No. 1 (1924).

Antić, Ljubomir. "Hrvatska federalistička seljačka stranka." *Radovi : Radovi Zavoda za hrvatsku povijest Filozofskoga fakulteta Sveučilišta u Zagrebu* 15/1 (1982), 136–222.

Arnaudov, Mihail. *Istoriia na sofiiskiia universitet sv. Kliment Ohridski prez purvoto mu polustoletie: 1888–1938.* Sofia: Pridvorna pechatnitsa, 1939.

Avramov, Rumen. *Komunalniat kapitalizum.* T.1-3. Sofia: Bulgarska nauka i kultura, Tsentur za liberalni strategii, 2007.

Avramović, Mihailo. *Trideset godina zadružnoga rada 1894–1924.* Belgrade: Zemunska štamparija glavnoga saveza srpskih zemljoradničkih zadruga, 1924.

Baeva, Iskra. "Agrarizmut kato ideologia na selskite dvizhenia v Bulgaria i Polsha do vtorata svetovna voina." *Godishnik na Sofiiskia Universitet "Sv.Kliment Ohridski," Istoricheski Fakultet* 84–85 (1992), 183–210.

Balcar, Jaromír. "Instrument im Volkstumskampf? Die Anfänge der Bodenreform in der Tschechoslowakei 1919/20." *Vierteljahrshefte für Zeitgeschichte* 46 (1998), 391–428.

Balibar, Étiene. "The Nation Form: History and Ideology." *New Left Review* XIII/3 (1990), 329–61.

Ballinger, Pamela. *History in Exile: Memory and Identity at the Borders of the Balkans.* Princeton, NJ: Princeton University Press, 2003.

Balkanski Institut. *Etat economique des pays balkaniques.* Vol 2: *Encyclopedie economique des Balkans.* Belgrad: Edition de l'Institut balkanique, 1938.

Baloutzova, Svetla. *Demography and the Nation: Social Legislation and Population Policy in Bulgaria, 1918–1944.* Budapest: Central European University Press, 2011.

Banac, Ivo. *The National Question in Yugoslavia: Origins, History, Politics.* Ithaca, NY: Cornell University Press, 1984.

Barev, Tsenko. *Prinos kum istoriiata na Bulgarskiia Zemedelski Naroden Suiuz.* Sofia: Bulvest 2000, 1994.

Baturinskii. "Agrarnie reformi v Evrope i ih osveschenie burzhuaznimi ekonomistami." *Na agrarnom fronte,* No. 1 (1926).

Becker, Annette, Stéphane Audoin-Rouzeau, and Leonard Smith, eds. *France and the Great War.* Cambridge: Cambridge University Press, 2003.

Bell, John. *Peasants in Power: Alexander Stamboliiski and the Bulgarian Agrarian National Union 1899–1923*. Princeton, NJ: Princeton University Press, 1977.

Berend, Ivan. "Agriculture in Eastern Europe: 1919–1939." In *Papers in East-European Economics*. Vol. 35. Oxford: Centre for Soviet and East European Studies, St. Anthony's College, July 1973.

Berend, Ivan. *Decades of Crisis: Central and Eastern Europe before World War II*. Berkeley: University of California Press, 1998.

Berend, Ivan, and G. Ranki. *East-Central Europe in the 19th and 20th Centuries*. Budapest: Columbia University Press, 1974.

Berger, Stefan, and Alexei Miller, eds. *Nationalizing Empires*. Budapest: CEU Press, 2015.

Bergholz, Max. *Violence as a Generative Force: Identity, Nationalism, and Memory in a Balkan Community*. Ithaca, NY: Cornell University Press, 2016.

Berindei, Dan, ed. *Der Bauer Mittel- und Osteuropas im sozio-ökonomischen Wandel des 18.und 19. Jahrhunderts. Beiträge zu seiner Lage und deren Widerspiegelung in der zeitgenössischen Publizistik und Literatur*. Köln; Wien: Böhlau, 1973.

Berov, L. "Sotsialnata struktura na seloto v balkanskite strani prez perioda mezhdu dvete svetovni voini." In *Trudove na Visshiia Ikonomicheski Institut «K.Marx» (4)*. Sofia: Nauka i izkustvo, 1977, 41–79.

Berov, Ljuben. "The Idea of a Cooperative Society in East European Peasant Movements During the Interwar Period." In Ferenc Glatz, ed. *Modern Age – Modern Historian. In Memoriam György Ránki (1930–1988)*. Budapest: Institute of History of the Hungarian Academy of Sciences, 1990, 265–86.

Bićanić. *Kako živi narod. Život u pasevnim krajevima*. Vol. 1. Zagreb: Nakladni zavod Globus, 1936.

Biondich, Mark. *Stjepan Radić, the Croat Peasant Party, and the Politics of Mass Mobilization, 1904–1928*. Toronto: University of Toronto Press, 2000.

Bizzell, William Bennett. *The Green Rising: An Historical Survey of Agrarianism, with Special Reference to the Organized Efforts of the Farmers of the United States to Improve Their Economic and Social Status*. New York: Macmillan, 1926.

Black, Cyril E. *Challenge in Eastern Europe: 12 Essays*. New Brunswick, NJ: Rutgers University Press, 1954.

Boban, Branka. *Demokratski nacionalizam Stjepana Radića*. Zagreb: Zavod za hvtsku povijest Filozofskog fakulteta Sveučilišta u Zagrebu, 1998.

Boban, Branka. "O osnovnim obilježnjima seljacke drzave u ideologije Antuna i Stjepana Radića." *Radovi Insituta za hrvatsku povijest*. Vol. 13. Zagreb, 1980.

Boban, Branka. "Stjepan Radič – opus, utjecaji i dodiri." *Radovi. Zavod za hrvatska povjest. Filosfskog fakulteta* 22 (Zagreb, 1989).

Boban, Ljubo. *Maček i politika Hrvatske seljačke stranke 1928–1941*. Zagreb: Liber, 1974.

Bogdanova, Rumiana. "Zemedelskoto dvizhenie sled vtorata svetovna voina (Putiat na partiinoto obezlichavane)." In Georgi Markov, ed. *Zemedelskoto dvizhenie v Bulgaria:*

*Istoriia, razvitie, lichnosti*. Pazardhik: Regionalen istoricheski muzei, Institut po istoriia pri BAN, 2004, 264–72.

Bogojević, B. "Agrarna reforma." In *Jubilarni zbornik zhivota i rada Srba Hrvata I Slovenaca 1.XII.1918–1.XII.1928*. Vol. 1. Beograd: Matica živih i mrtvih Srba, Hrvata i Slovenaca, 1928, 299–316.

Boiadzhieva, Elinka. "Zhultite paveta—keramika, estetika, simvolika." *Nauka* 21/5 (2011), 64–67.

Bokovoy, Melissa. *Peasants and Communists: Politics and Ideology in the Yugoslav Countryside, 1941–1953*. Pittsburgh, PA: University of Pittsburgh Press, 1998.

Bokovoy, Melissa. "Peasants and Partisans: A Dubious Alliance." In Norman Naimark and Leonid Gabianskii, eds. *The Establishment of Communist Regimes in Eastern Europe, 1944–1949*. Boulder, CO: Westview Press, 1997, 167–89.

Boll, Michael M. *Cold War in the Balkans: American Foreign Policy and the Emergence of Communist Bulgaria 1934–1947*. Lexington: University of Kentucky Press, 1984.

Bollmann, Kerstin. *Agrarpolitik. Entwicklungen und Wandlungen zwischen Mittelalter und Zweitem Weltkrieg*. Frankfurt (Main): Peter Lang, 1990.

Borras, Jr., Saturnino M., Marc Edelman, and Cristóbal Kay, eds. *Transnational Agrarian Movements Confronting Globalization*. Malden, MA: Wiley-Blackwell, 2008.

Botev, Sava, and Iosif Kovachev. *Zemledelieto v Bulgaria*. Sofia: Pechatnitsa S. M. Staikov, 1934.

Bozhinov, Aleksandur. *Album ot karikaturi iz nashiia obshtestven i politicheski zhivot*. Sofia: Pechatnitsa V. Poshta, 1907.

Bozhinov, Aleksandur. *Karikaturi*. Sofia: Izdava komiteta za chestvuvane 25-godishnata deinost na hudozhnika, 1924.

Bozhinov, Aleksandur. *Karikaturi i skitsi*. Sofia: Izdanie na Bulgarskata Akademia na naukite, 1957.

Bozhinov, Aleksandur. *Minali dni, izdanie dopulneno s nepublikuvani snimki*. Sofia: Iztok-Zapad, 2017.

Bozhinov, Aleksandur. *Za pruv put Aleksandur Bozhinov razkazva*. Sofia: Iztok-Zapad, 2015.

Brass, Tom. *Peasants, Populism and Postmodernism: The Return of the Agrarian Myth*. London: Frank Cass, 2000.

Brdlík, Vladislav. *Die sozialökonomische Struktur der Landwirtschaft in der Tschechoslowakei*. Berlin: Verlag Franz Vahlen, 1938.

Bratkov, Petur. "Selsko-stopanskata evoliutsiia i agrarnute krizi." *Arhiv za stopanska i sotsialna politika*, No. 2 and 3 (1929).

Brown, Carl L., ed. *Imperial Legacy: The Ottoman Imprint on the Balkans and the Middle East*. New York: Columbia University Press, 1996.

Brown, Keith. *The Past in Question. Modern Macedonia and the Uncertainties of Nation*. Princeton, NJ: Princeton University Press, 2003.

Brubaker, Rogers. *Nationalism Reframed: Nationhood and the National Question in the New Europe*. Cambridge: Cambridge University Press, 1996.

Brus, Wlodzimierz. *Geschichte der Wirtschaftspolitik in Osteuropa*. Köln: Bund-Verlag, 1986.

Brzeski, Andrzej. "Industrialization and Peasantry: An Economist View." *East European Quarterly* 3–4 (1970), 406–19.

Buchhofer, Ekkehard, ed. *Agrarwirtschaft und ländlicher Raum Ostmitteleuropas in der Transformation*. Marburg: Herder-Institut, 1998.

Bucur, Maria. *Heroes and Victims. Remembering War in Twentieth Century Romania*. Bloomington: Indiana University Press, 2009.

Bugajski, Janusz. *Political Parties of Eastern Europe: A Guide to Politics in the Post-Communist Era*. Armonk, NY: M. E. Sharpe, 2002.

*Bulgarski Periodichen Pechat 1844–1944*. Sofia: Nauka i Izkustvo, 1969.

Buxton, Noel. "The Balkans Today." *Nineteenth Century* XC (August, 1921), 333–35.

Cambel, Samuel. *Štátnik a národohospodár Milan Hodža, 1878–1944*, Bratsilava: VEDA, vydavateľstvo SAV, 2001.

Castoriadis, Cornelius. *The Imaginary Institution of Society* [*L'Institution imaginaire de la société*]. Cambridge, MA: MIT Press, 1987.

Chakrabarty, Dipesh. *Habitations of Modernity: Essays in the Wake of Subaltern Studies*. Chicago, IL: University of Chicago Press, 2002.

Chakrabarty, Dipesh. *Provincializing Europe*. Princeton, NJ: Princeton University Press, 2007.

Chary, Frederick B. *The History of Bulgaria*. Santa Barbara, CA: Greenwood, 2011.

Chatterjee, Partha. *The Nation and Its Fragments*. Princeton, NJ: Princeton University Press, 1993.

Chatterjee, Partha. "Subaltern History." In Neil Smelser and Paul Bates, eds. *International Encyclopedia of the Social and Behavioral Sciences*. Amsterdam: Elsevier Science, 2001, 15237–41.

Chatterjee, Partha. "Whose Imagined Community?" *Millenium: Journal of International Studies* 20/3 (Winter 1991), 521–25.

Chayanov, Alexander V. *The Theory of the Peasant Economy*. Edited by D. Thorner, B. Kerblay, and R. E. F. Smith. Homewood, IL: Richard D Irwin Inc for the American Economic Association Translation Series, 1966.

Chibber, Vivek. *Postcolonial Theory and the Specter of Capital*. London: Verso, 2013.

Cipek, Tihomir. *Ideja hrvtske države u političkoj misli Stjepana Radića*. Zagreb: Alenea, 2001.

Conacher, H. M. "La reforme agraire dans L'Europe orientale." *Revue Internationale des Institutions Economiques et Sociales*, No. 1 (1923).

Crampton, Richard. *Aleksandur Stamboliiski. Bulgaria*. London: House Histories, 2009.

Crampton, Richard. *Eastern Europe in the Twentieth Century–and After*. 2nd edition. New York: Routledge, 1997.

Crampton, Richard. "Modernization: Conscious, Unconscious and Irrational." In Ronald Schönfeld, ed. *Indistrialisierung und gesellschaftlicher Wandel in Südosteuropa*. München: Südosteuropa-Gesellschaft, 1989, 125–34.

Crampton, Richard. *A Short History of Modern Bulgaria*. Cambridge: Cambridge University Press, 1987.

Creed, Gerald. *Domesticating Revolution: From Socialist Reform to Ambivalent Transition in a Bulgarian Village*. University Park: Pennsylvania State University Press, 1998.

Crouthamel, Jason. *The Great War and German Memory: Society, Politics and Psychological Trauma, 1914–1945*. Liverpool: University of Liverpool Press, 2010.

Damaschke, Adolf. *Die Bodenreform. Grundsaetzliches und Geschichtliches zur Erkenntnis und Ueberwindung der sozialen Not*. Jena: Fischer, 1919.

Daskalov, Raiko. *Borba za Zemia*. Sofia: Pechatnitsa "Zemedelsko Zname," 1923.

Daskalov, Raiko. *Izbrani proizvedenia*. Edited by Svetla Daskalova and Dimitur Tishev. Vol. I–II. Sofia: Izdatelstvo na BZNS, 1986.

Daskalov, Roumen. *Debating the Past: Modern Bulgarian History from Stambolov to Zhivkov*. Budapest: CEU Press, 2011.

Daskalov, Roumen. *Ot Stambolov do Zhivkov. Golemite Sporove za Novata Bulgarska Istoriia*. Sofia: IK "Gutenberg," 2009.

Deak, Istvan, Jan Gross, and Tony Judt, eds. *The Politics of Retribution in Europe: World War II and Its Aftermath*. Princeton, NJ: Princeton University Press, 2000.

Dedjer, Vladimir, Ivan Božić, Sima Čirković, and Milorad Ekmečić. *History of Yugoslavia*. New York: McGraw-Hill, 1975.

Deianova, M. "Agrarni problemi v balkanskite strani." *Arhiv za stopanska i sotsialna politika* X/3 (1935).

Deianova, M. "Edin pogled vurhu zemedelskite usloviia." *Narodno Stopanstvo* XXXIX/7 (1933).

Deianova, M. "Osobenostite na zemedelskoto proizvodstvo i tiahnoto znachenie za agrarnite meropriatia." *Narodno Stopanstvo* XXXV/4 (1929).

Delev, N. S. "Stopanskoto preustroistvo na Evropa i zamedelskite strani ot Yugoiztoka." *Zemedelsko-stopanski vuprosi* VI/2 (1941).

Della Porta, Donatella, and Alberto Vannucci. *Corrupt Exchanges: Actors, Resources, and Mechanisms of Political Corruption*. New York: Aldine de Gruyter, 1999.

Desbons, G. *La Bulgarie apres le Traite de Neuilly*. Paris: Éditions M. Rivière, 1930.

Dichev, Petko. *Za bulgarskoto selsko stopanstvo i za kooperatsiata v 40 godini. Sbornik ot trudove 1890–1930*. Sofia: Kooperativna pechatnitsa "Gutenberg," 1933.

Dimitrov, George M. "Agrarianism." In Feliks Gross, ed. *European Ideologies*. New York: Philosophical Library, 1947, 396–450.

Dimitrov, Dr., G. M. *Prisuda sreshtu komunizma (1949/72): statii, rechi, izkazvania*. Sofia: Izdatelska kushta «Tsanko Tserkovski,» 1991.

Dimou, Augusta. *Entangled Paths Toward Modernity: Contextualizing Socialism and Nationalism in the Balkans*. Budapest: Central European University Press, 2009.

Dimov, Dimitur. *BZNS i voinishkoto vustanie*. Sofia: Izdatelstvo na BZNS, 1979.

Dimov, Nencho. "Agrarniiat vupros i selskoto dvizhenie v mezhduvoenna Albaniia." In Nikolai Todorov, ed. *Iunskata bourzh-dem revoliutsiia v Albaniia*. Sofia: Izdatelstvo na Bulgarskata Akademiia na naukite, 1989.

Dimov, Nencho. "Aleksandur Stamboliiski i selskite partii ot stranite na Yugoiztochna Evropa prez purvite desetiletiia na XX vek." *Studia Balkanica* V/19 (Sofia, 1986), 203-46.

Dimov, Nencho. "Krestinternut i selskite partii v stranite na Yugoiztochna Evropa (1923-29)." *Studia Balkanica* 17 (Sofia, 1983), 240-53.

Dimov, Nencho. *Profinternut ii balkanskoto profsuiuzno dvizhenie*. Sofia: Profizdat, 1976.

Dimov, Nencho. "V. Kolarov i mezhdunarodnoto agrarno dvizhenie (1929-39)." *Istoricheski Pregled* XXXIII/4 (1977), 3-26.

Dipper, Christof. "Bauern als Gegenstand der Sozialgeschichte." In Wolfgang Schieder and Volker Sellin eds. *Sozialgeschichte in Deutschland*. Göttingen: Vandenhoeck & Ruprecht, 1987, S9-33.

Djordjevic, Dimitrije. "Agrarian Reforms in Post World War One Balkans." *Balkanica*, Belgrade XIII-XIV (1982-1983), 255-69.

Dombal, G. "Agrarno-krestianskii vopros v Jugoslavii." *Agrarnye problemy* 1/3 (Moskva, 1928).

Doig, Alan, and Robin Theobald, eds. *Corruption and Democratization*. Portland, OR: Frank Cass, 2000.

Dolmanyos, Istvan. "Der Platz der Balkanischen Bodenreformen im Rahmen der Osteuropaeischen Agrarreformserie (1917-1939)." *Actes du premier congres international des etudes balkaniques*. Vol. 5. Sofia: Académie bulgare des sciences, 1970.

Dolmanyos, Istvan. "Le probleme des reformes agraires dans l'Europe Orientale apres la Premiere Guerre Mondiale." *Annales Universitatis Scientiarum Budapetienensis de Rolando Eotvos, Sectia Hist*. t. IV, Budapest, 1962.

Dostál, Vladimir. *Agrární strana. Její rozmach a zánik*. Brno: Atlantis, 1998.

Dostál, Vladimir. *Antonín Švehla. Profil československého státníka*. Praha: Státní Zemedelské Nakl., 1990.

Dubrovskii, S. M. "Sovremennii agrarnii vopros ii zadachi ego izucheniia." *Agrarnye Problemy* 1/1 (Moskva, 1927).

Eellend, Johan. "Agrarianism and Modernization in Inter-War Eastern Europe." In Piotr Wawrzeniuk, ed. *Societal Change and Ideological Formation among the Rural Population of the Baltic Area 1880-1939*. Huddinge: Södertörns högskola, 2008.

Eisenstadt, S. N. "Multiple Modernities." *Daedalus* 129/1 (Winter 2000), 1-30.

*Enquete sur les mouvements paysans dans le monde contemporain (de la fin du 18 siecle a nos jours)*. XIII Congres International des sciences historiques. Moscou, 16-23 aout 1970.

Erić, Milivoje. *Agrarna reforma u Jugoslaviji 1918-41*. Sarajevo: Veselin Masleša, 1958.

Evelpidis, Ch. *Les Etats balkaniques. Etude comparee, politique, sociale, economique et financiere*. Paris: Rousseau, 1930.

Fellner, Fritz. "The Dissolution of the Habsburg Monarchy and Its Significance for the New Order in Central Europe: A Reappraisal." *Austrian History Yearbook* 4 (1968), 3-27.

Fellows, J. W. *Antecedents of the International Labour Organization*. Oxford: Clarendon Press, 1951.

Fischer-Galati, Stephen. "Peasantism in Interwar Eastern Europe." *Balkan Studies* 8 (1967), 103–14.

Freidzon, V. I. "Sotsialno-politicheskie bzgliadi Antuna i Stepana Radichei v 1900-x godah i voznaknovenie Horvatskoi krestianskoi partii (1904–1905)." *Uchenie zapiskiistituta slavianovedeniia*, G. 20 (Moskva, 1960), 275–305.

Fussel, Paul. *The Great War and Modern Memory*. Oxford: Oxford University Press, 2000.

Gaković, Milan. *Savez Zemljoradnika (Zemljoradnička Stranka 1919–1941)*. Priredio Zdravko Antonić. Banja Luka: Akademija Nauka i Umjetnosti Republike Srpske, 2008.

Ganev, Ivan. *Agrarnata reforma v chuzhbina i u nas*. Sofia: Pechatnitsa "G. Provadaliev," 1946.

Gargas, Sigismund. *Die grüne Internationale*. Halberstadt: H. Meyer's Buchdruck, 1927.

Gellner, Ernest. *Nations and Nationalism*. Ithaca, NY: Cornell University Press, 1983.

Genovski, Mihail. *Aleksandur Stamboliiski otblizo i otdaleko*. Sofia: Izdatelstvo na BZNS, 1982.

Georgieva, Elena, S. Zherev, and V. Stankov, eds. *Istoriia na novobulgarskiia knizhoven ezik*. Sofia: Izdatelstvo na Bulgarskata Akademiia na Naukite, 1989.

Geshkoff, Theodore. *The Balkan Union: A Road to Peace in Southeastern Europe*. New York: Columbia University Press, 1940.

Glad, Betty, and Eric Shiraev, eds. *The Russian Transformation: Political, Sociological, and Psychological Aspects*. New York: St. Martin's Press, 1999.

Gligorijević, Br. *Demokratska stranka I politicki odnosi u Kraljevini Srba, Hrvata i Slovenaca*. Beograd: Institut za savremenu istoriju, 1970.

Gligorijević, Br. "Parlamentarni sistem u Kraljevi SHS. 1919–1929." In Aleksandar Acković, ed. *Politicki zivot Jugoslavije. 1914–1945: Zbornik radova*. Beograd: Radio-Beograd, 1973.

Gligorievič, Br. *Parlament i politicke stranke u Jugoslaviji. 1919–1929*. Beograd: Institut za savremenu istoriju, 1979.

Gollwitzer, Heinz, ed. *Europäische Bauernparteien im 20. Jahrhundert*. Stuttgart: Gustav Fischer Verlag, 1977.

Goranovich, M. "Profesionalna struktura na naselenieto v Yugoslavia." *Arhiv za stopanska i sotsialna politika*, God. XIV, kn. 3 (1939).

Goranovich, M. M. *Agrarnii krizis i razpad agrarnogo bloka stran Vostochnoi u Iugo-vostochnoi Evropi. 1930–1933*. Moskva: Nauka, 1971.

Goranovich, M. M. *Krah zelenogo internatsionala*. Moskva: Nauka, 1967.

Gorov, M. "Agrarnii krestianskii vopros i Komintern." *Agrarnye Probemyl*, No. 4 and 5 (1930).

Gorova, ed. *Agrarnii vopros i krestianskoe dvizhenie*. Spravochnik, Moskva: Mezhdunarodnyi agrarnyi institut, 1936.

Grebenarov, Alexander. "VMRO i SSSR—Sblizhenie s predizvesten krai (1923–1924)." In *Mezhdunarodna Nauchna Konferentsiia 'Bulgariia I Rusiia mezhdu priznatelnostta i pragnatizma—Dokladi*. Sofia, March 2008.

Greenfeld, Leah. *Nationalism: Five Roads to Modernity*. Cambridge, MA: Harvard University Press, 1992.

Gross, Feliks, ed. *European Ideologies: A Survey of 20th Century Political Ideas*. Freeport, NY: Books for Libraries Press, 1948.

Gross, Jan. "War as Revolution." In Norman Naimark and Leonid Gabianskii, eds. *The Establishment of Communist Regimes in Eastern Europe, 1944–1949*. Boulder, CO: Westview Press, 1997, 17–40.

Gunchev, Zdravko. *Razvitie na ideate za kooperirane na zemedelskite stopani v Bulgaria*. Varna: Nauka i izkustvo, 1949.

Gunev, Georgi. *Kum brega na svobodata ili za Nikola Petkov i negovoto vreme*. Sofia: Infornatsionno obsluzhvane AD, 1989, 1992.

Gurdev, Konstantin. "Bulgarskata zemedelska politicheska emigratsiia v Chehoslovakia prez 20-te i 30-te godini na XX vek." In Milen Kumanov and Rumiana Katsarova, eds. *Zemedelskoto dvizhenie v Bulgaria: istoriia, razvitie, lichnosti*. Pazardzhik: Belloprint, 2004, 213–17.

Gyllin, Roger. *The Genesis of the Modern Bulgarian Literary Language*. Stockholm: Uppsala, 1991.

Hall, Richard C. *Balkan Breakthrough: The Battle of Dobro Pole, 1918*. Bloomington: Indiana University Press, 2010.

Haller, Dieter, and Cris Shore. *Corruption: Anthropological Perspectives*. Ann Arbor, MI: Pluto Press, 2005.

Hagen, Mark von, and Karen Barkey, eds. *After Empire: Multiethnic Societies and Nation-building: The Soviet Union and Russian, Ottoman, and Habsburg Empires*. Boulder, CO: Westview Press, 1997.

Hatzopoulos, Pavlos. *The Balkans beyond Nationalism and Identity: International Relations and Ideology*. London: Tauris, 2008.

Hayes, Carleton. *The Historical Evolution of Modern Nationalism*. New York: MacMillan, 1931, 1948.

Herceg, Rudolf. *Die Ideologie der kroatischen Bauernbewegung. Mit einem Vorwort von Stjepan Radic*. Zagreb: Herceg, 1923.

Heumos, P. "Agrarische Organisationen und nationale Mobilisierung in Böhmen im 19. Jahrhundert: ein Überblick." In W. Conze, G. Schram, and K. Zernack, eds. *Modernisierung und nationale Gesellschaft im ausgehenden 18. und im 19. Jahrhundert: Referate einer deutsch-polnischen Historikerkonferenz*. Berlin: Duncker & Humblot, 1978.

Heumos, Peter. "Die Entwicklung organisierter agrarischer Interessen in den böhmischen Ländern und in der ČSR. Zur Entstehung und Machtstellung der Agrarpartei 1873–1938." In Karl Bosl, ed. *Die Erste Tschechoslowakische Republik als multinationaler Parteienstaat*. München; Wien, 1979, 323–76.

Hilendarski, Paisii, and Petur Dinekov, ed. *Slavianobulgarska istoriia*. Sofia: Bulgarski Pisatel, 1972.

Hitchins, Keith. "A Rural Utopia. Virgil Madgearu and Peasantism." In Keith Hitchins. *The Identity of Romania*. Bucharest: Encyclopaedic Publishing House, 2003, 151–203.

Hobson, Asher. *The International Institute of Agriculture; an Historical and Critical Analysis of Its Organization, Activities and Policies of Administration*. Berkeley: University of California Press, 1931.

Holmes, Leslie. *The End of Communist Power: Anti-corruption Campaigns and Legitimation Crisis*. New York: Oxford University Press, 1993.

Holmes, Leslie. *Rotten States? Corruption, Post-Communism, and Neoliberalism*. Durham, NC: Duke University Press, 2006.

Horvat, Josip. *Politička povijest Hrvatske*. Zagreb: "August Cesarec," 1990.

Houee, P. *Les Etapes du developpement rural*. Paris: Editions Economie et Humanisme, 1972.

Houillier, F. *L'organisation internationale de l'agriculture. Les institutions agricoles internat et l'action internat en agriculture*. Paris, 1935.

Hristov, Hristo. *Revoliutsionnata kriza v Bulgariia prez 1918–1919*. Sofia: Izdatelstvo na Bulgarskata Komunisticheska Partiia, 1957.

Hroch, Miroslav. *Social Preconditions of National Revival in Europe: A Comparative Analysis of the Social Composition of Patriotic Groups among the Smaller European Nations*. Cambridge: Cambridge University Press, 1985.

Ianulov, Ilia. *Mezhdunarodnata organizatsia na truda 1919–1938*. Sofia, 1938.

Ianulov, Ilia. "Stopanskoto i sotsialno polozhenie na Bulgariia i razgovora za mir." *Arhiv za stopanska i sotsialna politika*. kn. 2,3 (1929).

Iarkin, Iu. "K agrarnoi reforme v Yugoslavii." *Agrarnye problemy*, No. 3–4 (Moskva, 1927).

Ionescu, Ghita, and Ernest Gellner. *Populism: Its Meaning and National Characteristics*. New York: Macmillan, 1969.

Iosifov, Kalin. *Totalitarnoto nasilie v bulgarskoto selo (1944–1951) i posleditisite za Bulgaria*. Sofia: Universitetsko Izdatelstvo "Kliment Ohridski," 1999, 2003.

Isić, Momčilo. *Seljaštvo u Srbiji 1918–1941*. Belgrad: Institut za noviju istoriju Srbije, 1995.

Isić, Momčilo. *Socijalna i agragna struktura Srbje u kraljevini Jugoslaviji (prema popisu stanvnishtva od 31. Marta 1931 godine)*. Beograd: Institu za noviju istoriju Srbje, 1999.

Isić, Momčilo, ed. *Svileuva*. Godishnjak 1 (2003), 2 (2004), 3 (2005), 4 (2006), 5 (2007). Svilueva: Drushtvo za izuchavanje istorije Svilueve.

*Istoriia na Bulgariia v chetirinadeset toma*, t.5, Sofia: Izdatelstvovo na Bulgarskata akademiia na naukite, 1984.

*Istoriia srpskog naroda*. 6th vol. 1878–1918. Beograd, 1983.

Ivanov, Vulo. *Borbata protiv fashizma za demokratsia v Narodnoto subranie, 1923–1944*. Sofia: Polimona, 1997.

Ivanov, Vulo. *Protiv fasjizma. Sbornik statii. 1923-1944*. Sofia: M&M, 1993.
Ivsic, M. *Les problemes agraires en Yugoslavie*. Paris: Librairie Artur Rousseau, 1926.
Jackson, Jr., George D. *Comintern and Peasant in East Europe*. New York: Columbia University Press, 1966.
Jain, Arvind K. *The Political Economy of Corruption*. New York: Routledge, 2001.
Janjatović, Bosiljka. *Stjepan Radić: Progoni - Zatvori - Suđenja - Ubojstvo, 1889-1928*. Zagreb: Dom i Svijet, 2003.
Jankovic, Dragoslav. *Stvaranje Jugoslavije*. Beograd: Visoka škola političkih nauka, 1961.
Jareb, Jere. *Pola stoljeća hrvatske politike*. Zagreb: Institut za suvremenu povijest, 1995.
Jelavich, Barbara. *History of the Balkans: Twentieth Century*. Vol. 2. Cambridge: Cambridge University Press, 1983.
Jelić-Butić, Fikreta. *Hrvatska Seljačka Stranka*. Zagreb: Globus, 1983.
Jelić-Butić, Fikreta. *Ustaše i Nezavisna Država Hrvatska 1941-1945*. Zagreb: Liber, 1977.
Jordan, Constantin. "Les reformes agraires dans la periode de l'entre deux guerres: Repers comparatifs." *Revue des etudes sud-est europeennes*, No. 3 (Buchgarest, 1984), 243-54.
Jovanović, Jovan. *Diplomatska istorija nove Evrope 1918-1938*. Kniga II. Beograd: Koste Mihailović, 1939.
Jovanović, Nadežda. *Zemljoradnička levica u Srbiji 1927-1939*. Beograd: INIS, 1994.
Jovanović, Nadežda. *Život za svobodu bez straha*. Beograd: INIS, 2000.
Jovanovic, Z. *KPJ prema seljastvu. 1919-1941*. Beograd: Narodna Knjiga, 1984.
Judt, Tony. *Postwar: A History of Europe since 1945*. New York: Penguin, 2006.
Kalinova, Evgenia, and Iskra Baeva. *Bulgarskite prehodi 1939-2005*. Sofia: Paradigma, 2005.
Karakasidou, Anastasia. *Fields of Wheat, Hills of Blood: Passages to Nationhood in Greek Macedonia, 1870-1990*. Chicago, IL: University of Chicago Press, 1997.
Karklins, Rashma. *The System Made Me Do It: Corruption in Post-Communist Societies*. Armonk: M. E. Sharpe, 2005.
Kaser, Michael C. *Economic Development for Eastern Europe. Proceedings of a Conference Held by the International Econ. Assopciation*. London: Macmillan, 1968.
Kaser, Michael C., and E. A. Radice. *The Economic History of Eastern Europe, 1919-1975*. Oxford: Clarendon Press, 1985.
Katsarkova, Vera, and Echka Damianova. "Agrarniiat protektsionism v balkanskite strani mezhdu dvete svetovni voini." In *Protektsii i konkurentsiia na Balkanite*. Sofia: Izdatelstvo an Bulgarskata Akademiia na Naukite, 1989.
Kerner, Robert Joseph, and Harry Nocholas Howard. *The Balkan Conferences and the Balkan Entente 1930-1935: A Study in Recent History of the Balkans and the near Eastern Peoples*. Westport, CT: Greenwood Press, 1936.
King, Jeremy. "The Natonalization of East Central Europe: Ethnicism, Ethnicity, and Beyond." In Maria Bucur and Nancy M. Wingfield, eds. *Staging the Past: the Politics of Commemoration in Habsburg Central Europe, 1848 to the Present*. West Lafayette, IN: Purdue University Press, 2001, 112-52.

Kligman, Gail, and Katherine Verdery. *Peasants under Siege: The Collectivization of Romanian Agriculture, 1949–1962*. Princeton, NJ: Princeton University Press, 2011.

Kohn, Hans. *Prophets and Peoples: Studies in Nineteenth Century Nationalism*. New York: Macmillan, 1946.

Kolář, Josef. *Bulharská demokratická emigrace v Československu v letech 1923–1933*. Prague: Academia, 1983.

Kolář, Josef. "Chehoslovashkata diplomatsia i pravitelstvoto na Bulgarskiia Zemedelski Naroden Suyuz." In *120 godini ot rozhdenieto na Aleksandur Stamboliiski*, Sofia: Anubis, 1999.

Kolarov, Vasil. "Krestianskie partii i soiuzi." *Kommunisticheskii Internatsional* 7/3 (1925), 3–20.

Kubů, Eduard, ed. *Mýtus a realita hospodárské vyspelosti Ceskoslovenska mezi svetovými válkami*. Praha: Nakl. Karolinum, 2000.

Kubů, Eduard, Torsten Lorenz, Uwe Müller, and Jiří Šouša, eds. *Agrarismus und Agrareliten in Ostmitteleuropa*. Berlin-Praha: Berliner Wissenschaftsverlag – Dokořán, 2013.

Kumanov, Milen, and Rumiana Katsarova, eds. *Zemedelskoto dvizhenie v Bulgaria: istoriia, razvitie, lichnosti*. Pazardzhik: Belloprint, 2004.

Labrousse, Ernest, ed. *Mouvements nationaux d'indépendance et classes populaires aux XIXe et XXe siècles, en Occident et en Orient*. Vol. 1–2. Paris: Librairie Armand Colin, 1971.

Lampe, John, and Marvin Jckson. *Balkan Economic History. 1550–1950. From Imperial Borderlands to Developing Nations*. Bloomington: Indiana University Press, 1982.

Lazard, Max. *Compulsory Labor Service in Bulgaria*. Geneva: International Labour Office, 1922.

Leff, Nathaniel. "Economic Development through Bureaucratic Corruption." *American Behavioral Scientist* 8/3 (November 1964), 8–14.

Livezeanu, Irina. *Cultural Politics in Greater Romania: Regionalism, Nation Building, and Ethnic Struggle, 1918–1930*. Ithaca, NY: Cornell University Press, 1995.

Livingstone, G. "Stepan Radic and the Croatian Peasant Party. 1904 – 1929." Unpublished PhD dissertation, Department of History, Harvard University Press, 1959.

Lukáč, Pavol. *Milan Hodža v zápase o budúcnosť strednej Európy v rokoch 1939–1944*. Edited by Štefan Šebesta. Bratislava: VEDA, Vydateľstvo SAV, 2005.

Maček, Vladko. *In the Struggle for Freedom*. New York: Pennsylvania State University Press, 1957.

Madgearu, Vergil N. *L'Europe agricole devant la societe nations*, 1930.

Maier, Charles. *Recasting Bourgeois Europe*. Princeton, NJ: Princeton University Press, 1975.

Manchev, Krustiu. "Burzhoaznata demokratsiia na Balkanite mezhdu dvete svetovni voini." *Balkanistika* 2 (Sofia, 1987).

Marcheva, Iliana. "Zemedelskoto dvizhenie i negovata sudba v politicheskata sistema na Bulgariia sled 1989 godina." In Milen Kumanov and Rumiana Katsarova, eds.

*Zemedelskoto dvizhenie v Bulgaria: istoriia, razvitie, lichnosti*. Pazardzhik: Belloprint, 2004, 323-33.

Maruna, V. "Agrarno-krestianskii vopros v politike kommunisticheski partii v stran Tsentralnoi i Yugovostochnoi Evropy mezhdu dvumia mirovymi voinami." *Sovetskoe slavianovedenie* 10/4 (1974), 23-40.

Matković, Hrvoje. *Povijest Hrvatske Seljačke Stranke*. Zagreb: Naklada Pavičić, 1999.

Matković, Hrvoje. *Suvremena politička povijest Hrvatske*. Zagreb: MUP RH; Policijska akademija, 1999.

Matl, J. "Historische Grundlagen der agrarsozialen Verhältnisse auf dem Balkan." *Vierteljahrschrift für Sozial- und Wirtschaftsgeschichte* 65/2 (1965), 145-67.

Matl, Josef. "Bauer und Grundherr in der Geschichte der Balkanvölker." In Franz Ronneberger and Gerhard Teich, eds. *Von der Agrar- zur Industriegesellschaft. Sozialer Wandel auf dem Lande in Südosteuropa*. Darmstadt: Hoppenstedt, 1969, 1-27.

Mény, Yves. "'Fin de siècle' Corruption: Change, Crisis and Shifting Values." *International Social Science Journal* 48/3 (1996), 309-20.

Meshteriakov, N. L. *Za roliata na seliachestvoto v obshtestvenite borbi*. Sofia: Ekzarh Iosif, 1930.

Mihutina, I. V. "O meste krestianskih partii v politicheskoi evoliutsii stran Tsentralnoi i Iugovostochnoi Evropi mezhdu pervoi i vtoroi mirovymi voinami." *Sovetskoe Slavianovedenie* 18/2 (1982), 23-42.

Mihutina, I. V. "O sotsialnom haraktere i sotsialnih aspektah programm i deiatelnosti krestianskih partii v stran Tsentralnoi i Iugovostochnoi Evropi." In Aleksandr Klevanskii, ed. *Sotsialnaia struktura i politicheskie dvizheniia v stranah Tsentralnoi i Iugovostochnoi Evropi: mezhvoennii period*. Moskva: Nauka, 1986, 152-218.

Miller, Daniel E. *Forging Political Compromise: Antonín Švehla and the Czechoslovak Republican Party 1918-1933*. Pittsburgh, PA: University of Pittsburgh Press, 1999.

Miller, Ruth A. *The Erotics of Corruption: Law, Scandal, and Political Perversion*. Albany: State University of New York Press, 2008.

Miller, Alexei, and Alfred J. Rieber, eds. *Imperial Rule*. Budapest: CEU Press, 2004.

Mirkovic, M. *Ekonomska historija Jugoslavije*. Zagreb: Ekonomski pregled, 1958.

Mitrany, David. *The Agrarian Foes of Bolshevism*. New York: Columbia University Press, 1958.

Mitrany, David. *The Land and Peasant in Roumania: The War and Agrarian Reform*. Oxford: Oxford University Press, 1930.

Mitrany, David. *Marx against the Peasant: A Study in Social Dogmatism*. Chapel Hill: University of North Carolina Press, 1952.

Mladenatz, Gromoslav. *Histoire des doctrines coopératives*. Paris: Presse Universitaire 1933.

Moritsch, Andreas. "Die Bauernparteien bei den Kroaten, Serben und Slowenen." In Heinz Gollwitzer, ed. *Europäische Bauernparteien im 20. Jahrhundert*. Stuttgart; New York: Fischer, 1977, 359-402.

Moser, Charles. *Dimitrov of Bulgaria. A Political Biography of Dr. G. M. Dimtrov.* Thornwood, NY: Caroline House Publishers, 1979.

Müller, Dietmar. *Agrarpopulismus in Rumänien. Programmatik und Regierungspraxis der Bauernpartei und der Nationalbäuerlichen Partei Rumäniens in der Zwischenkriegszeit.* Sankt Augustin: Gardez!-Verlag, 2001.

Murgescu, C., and D. Hureanu. *La reforme agraire en Europe apres la Premiere Guerre mondiale.* XV Congres international des sciences historiques. Bucharest, 1980. Acte II, III, IV.

Musil, Jiří, ed. *The End of Czechoslovakia.* Budapest and New York: Central European University Press, 1995.

Mužić, Ivan. *Stjepan Radić u Kraljevini Srba, Hrvata i Slovenaca.* Ljubljana; Zagreb: Hrvatsko književno društvo sv. Ćirila i Metoda, 1987.

Naimark, Norman, and Leonid Gabianskii, eds. *The Establishment of Communist Regimes in Eastern Europe, 1944–1949.* Boulder, CO: Westview Press, 1997.

Nairn, Tom. *The Break-up of Britain: Crisis and Neo-Nationalism.* London: Verso, 1981.

Naumov, Georgi, ed. *Istoriia na Sofiiskia universitet "Kliment Ohridski."* Sofia: Universitetsko Izdatelstvo "Kliment Ohridski," 1988.

Neuburger, Mary. *Balkan Smoke: Tobacco and the Making of Modern Bulgaria.* Ithaca, NY: Cornell University Press, 2013.

Nikolaeva, Aneliia, and Ruzha Marinska, eds. *Aleksandur Bozhinov, 1878–1968.* Sofia: Natsionalna Hudozhestvena Galeriia, 1999.

Ognianov, Ljubomir. *Bulgarskiiat Zemedelski Naroden Sujuz 1899–1912.* Sofia: Izdatelstvo "Kliment Ohridski," 1990.

Ogniianov, Liubomir. *Voinishkoto vustanie: 1918.* Sofia: Nauka i Izkustvo, 1978.

Omelianov, A. "A Bulgarian Experiment." In P. A. Sorokin, ed. *A Systematic Source Book in Rural Sociology.* Vol. II. Minneapolis: University of Minnesota Press, 1931, 638–47.

Oren, Nissan. *Revolution Administered: Agrarianism and Communism in Bulgaria.* Baltimore, MD; London: Johns Hopkins University Press, 1973.

Özkırımlı, Umut. *Theories of Nationalism: A Critical Introduction.* New York: Palgrave, 2000.

Panaiotov, Panaiot. "Borbata na samostoiatelnoto pravitelstvo na BZNS za po-spravedlivo razreshavane na materialnite i finansovite sanktsii na Nioiskiia miren dogovor (1920–1923)." In Konstantin Kosev, Vitka Toshkova, and Panaiot Panaiotov, eds. *Problemut za mira v mezhdunarodnite otnosheniia 1919–1980.* Sofia: Izdatelstvo na Bulgarskata Akademiia na Naukite, 1987, 7–63.

Panizza, Francisco, ed. *Populism and the Mirror of Democracy.* New York: Verso, 2005.

Parma, Miloš, ed. *Agenti Zelené Internacionály: nepřátelé n aši vesnice.* Prague: Orbis, 1952.

Paxton, Robert. *French Peasant Fascism: Henry Dorgères' Greenshirts and the Crises of French Agriculture, 1929–1939.* New York: Oxford University Press, 1997.

Peknik, Miroslav, ed. *Milan Hodža a integrácia strednej Európy.* Bratislava: VEDA Vydateslstvo Slovenskej akadémie vied, 2006.

Peknik, Miroslav, ed. *Milan Hodža: Statesman and Politician*. Bratislava: VEDA Publishing House of the Slovak Academy of Sciences, 2007.

Peknik, Miroslav, ed. *Milan Hodža: politik a žurnalista*. Bratislava: VEDA Vydateslstvo Slovenskej akadémie vied, 2008.

Perović, Latinka. *Srpski socijalisti 19. veka. Prilog istoriji socialističke misli*. Beograd: Izdavačka organizacija "Rad," 1985.

Perroux, Francois. *Les Reformes agraires en Europe*. 2 vols. Paris: Domat-Montchrestien, 1935.

Petkov, Nikola. *Aleksandur Stamboliiski: lichnost i idei*. Sofia: Pechatnitsa "S.T. Charukchiev," 1930, 1946.

Petrone, Karen. *The Great War in Russian Memory*. Bloomington: Indiana University Press, 2011.

Petrova, Dimitrina. *Aleksandur Stamboliiski—durzhavnikut reformator*. Stara Zagora: Izdatelstvo "Znanie," 1995.

Petrova, Dimitrina. *Bulgarskiiat Zemedelski Naroden Suiuz, 1899–1944*. Sofia: Fondatsiia "Detelina," 1999.

Petrova, Dimitrina. *Samostoiatelnoto upravlenie na BZNS, 1920–1923*. Sofia: Durzhavno Izdatelstvo Nauka i Izkustvo, 1988.

Pisarev, Iu. A. "Polozhenie krestianstva v Serbo-harvato-slovenskom gosudarstve i agrarnaia reforma 1919." In P. N. Tretiakov, ed. *Uchenie zapiski instituta slavianovedenie*. T.4. Moskva: Izdatelstvo Akademii Nauk SSSR, 1951, 115–89.

Pisarev, Iu. A. "Iz istorii revoliutsionnogo dvizhenia rabochego klassa i krestianstva v Serbo-harvato-slovenskom gosudarstve v 1919–1923." In P. N. Tretiakov, ed. *Uchenie zapiski instituta slavianovedenie*. T.5. Moskva: Izdatelstvo Akademii Nauk SSSR, 1952, 78–150.

Popescu, Cornel, and George Daniel Ungureanu. "Romanian Peasantry and Bulgarian Agrarianism in the Interwar Period: Benchmarks for a Comparative Analysis." *The Romanian Review of Social Sciences* 4/16 (2014), 31–59.

Popova, Venche, ed. *Problemi ot istoriiata na bulgarskiia knizhoven ezik*. Sofia: Sofiiski Universitet "Kliment Ohridski," 1976.

Popov, Zhivko. *Ubitite zaradi ideite si. Politicheskiat vuzhod i zhiteiskoto krushenie na familia Petkovi*. Sofia: Kira 21, 2004.

Popovic, R. V. *Agrarna npenaseljenost Jugoslavije*. Doktorska razprava. Beograd, 1940.

Popovici, Aurel. *Die Vereinigten Staaten von Groß-Österreich. Politische Studien zur Lösung der nationalen Fragen und staatrechtlichen Krisen in Österreich-Ungarn*. Leipzig: Elischer, 1906.

Poppetrov, Nikolai. *Fashizmut v Bulgaria: Razvitie i proiavi*. Sofia: Kama, 2008.

Poppetrov, Nikolai. *Sotsialno naliavo, natsionalizmut—napred: Programni i organizatsionni dokumenti na bulgarski avtoritaristki natsionalisticheski formatsii*. Sofia: IK Gutenberg, 2009.

Prado, Italo. *Between Morality and the Law: Corruption, Anthropology and Comparative Society*. Aldershot: Ashgate, 2004.

Puhle, Hand-Jürgen. *Von der Agrarkrise zum Präfaschismus.* Wiesbaden: F. Steiner, 1972.

Puhle, Hans-Jürgen. *Politische Agrarbewegungen in kapitalistischen Industriegesellschaften. Deutschland, die USA und Frankreich im 20. Jahrhundert.* Göttingen: Vandenhoeck & Ruprecht, 1975.

Puhle, Hans-Jürgen. "Populismus, Krise und New Deal. Zum Verhältnis von agrarischerDemokratie und organisiertem Subventionismus in der Zwischenkriegszeit." In Heinrich August Winkler, ed. *Die große Krise in Amerika.* Göttingen: Vanderhoeck & Ruprecht, 1973, 107–52.

Puhle, Hans-Jürgen. "Warum gibt es in Westeuropa keine Bauernparteien?" In Heinz Gollwitzer, ed. *Europäische Bauernparteien im 20. Jahrhundert.* Stuttgart; New York: Fischer 1977, 603–67.

Puhle, Hans-Jürgen. "Zwischen Protest und Politikstil: Populismus, Neo-Populismus und Demokratie." In Nikolaus Werz, ed. *Populismus. Populisten in Übersee und Europa.* Opladen: Leske & Budrich, 2003, 15–43.

Pundeff, Marin. "The Bulgarian Academy of Sciences (On the Occasion of Its Centennial)." *East European Quarterly* III/3 (1969), 371–86.

Pundeff, Marin. "The University of Sofia at Eighty." *Slavic Review* 27/3 (September 1968), 438–46.

Rabinowitch, Alexander. *The Bolsheviks Come to Power: The Revolution of 1917 in Petrograd.* Chicago, IL: Haymarket Books, 2004.

Radelić, Zdenka. *Hrvatska Seljačka Stranka (1941–1950).* Zagreb: Hrvatski institut za povijest, 1996.

Radić, Stjepan. *Politički spisi, govori i dokumenti.* Zagreb: Dom i svijet, 1995.

Radulov, Stefan. "Ideinite osnovi na zemedelskia rezhim." In Akademia za obshtestveni nauki i sotsialno upravlenie na TsK na BKP. *Nauchni trudove.* Seria istoriia. Vol. 105. Sofia, 1979, 7–76.

Raikin, Spas. *Politicheski problemi pred bulgarskata obshtestvenost v chuzhbina: sbornik statii, eseta, studii i komentari po politicheski, kulturni, ikonomicheski i istoricheski vŭprosi, publikuvani v emigratsia 1978–1991.* Vol. 1. Veliko Turnovo: Pechat DF "Abagar," 1993.

Ramet, Sabrina P. *The Three Yugoslavias. State-Building and Legitimation, 1918–2005.* Bloomington: Indiana University Press, 2006.

Rankoff, Iwan. "Bauerndemokratie in Bulgarien." In Heinz Gollwitzer, ed. *EuropäischeBauernparteien im 20. Jahrhundert.* Stuttgart; New York: Fischer, 1977, 466–507.

*Rapport Mondial sur la Corruption 2004: Thème special: la corruption politique.* Paris: Éditions Karthala, 2004.

Reviakina, Luiza. *Kominternut i Selskite Partii na Balkanite: 1923–1931.* Sofia: Akademichno Izdatelstvo "Prof. M. Drinov," 2003.

Roberts, H. L. *Rumania: Political Problems of an Agrarian State.* New Haven, CT: Yale University Press, 1951.

Rose-Ackerman, Susan. *Corruption: A Study in Political Economy*. New York: Academic Press, 1978.
Rosenberg, Tina. *The Haunted Land: Facing Europe's Ghosts after Communism*. New York: Vintage Books, 1996.
Rösener, Werner. *The Peasantry of Europe*. Oxford: Blackwell, 1994.
Rothschild, Joseph. *The Communist Party of Bulgaria: Origins and Development, 1883–1936*. New York: Columbia University Press, 1959.
Rothschild, Joseph. *East Central Europe between the Two World War*. Seattle: University of Washington Press, 1974.
Rusinov, Rusin. *Istoriia na bulgarskiia pravopis*. Sofia: Nauka i Izkustvo, 1985.
Rusinov, Rusin. *Istoriia na novobulgarskiia knizhoven ezik*. Veliko Turnovo: Pechat "Abagar," 1999.
Rusinov, Rusin. *Uchebnik po istoriia na novobulgarskiia knizhoven ezik*. Sofia: Nauka i Izkustvo, 1980.
Sagadin, St. "Drzhavna uprava." *Jubilarni zbornik zhivota i rada SHS* 1.XII.1918–1.XII.1928 (Beograd, 1928).
Šarinić, Ivo. *Ideologija Hrvatskog seljačkog pokreta*. Zagreb: Albrecht, 1935.
Sarkar, Sumit. "The Decline of the Subaltern in *Subaltern Studies*." In Sumit Sarkar, ed. *Writing Social History*. Delhi: Oxford University Press, 1997, 82–108.
Schmidt, Amy K. *The Croatian Peasant Party in Yugoslav Politics*. Dissertation at Kent State University, 1984.
Schultz, Helga, and Angela Harre, eds. *Bauerngesellschaften auf dem Weg in die Moderne. Agrarismus in Ostmitteleuropa 1880 bis 1960*. Wiesbaden: Harrassowitz, 2010.
Scott, James. *Comparative Political Corruption*. Englewood Cliffs, NJ: Prentice-Hall, 1972.
Scott, James C. *The Moral Economy of the Peasant: Rebellion and Subsistence in Southeast Asia*. New Haven, CT: Yale University Press, 1976.
Šebesta, Štefan, ed. (Dissertation of Pavol Lukáč). *Milan Hodža v zápase o budúcnosťstrednej Európy v rokoch 1939–1944*. Bratislava: VEDA Vydateslstvo Slovenskej akadémie vied, 2005.
Semerdzhiev, Petur. *Sudebniat protses sreshtu Nikola Petkov*. Sofia: n.p., 1990.
Seton-Watson, H. *Eastern Europe between the Wars*. Cambridge: Harper & Row, 1962.
Shafir, I. A. "Agrarnii protektsionizum i sotsial'naia demokratiia." *Agrarnye problemy* (1928), 2.
Shafir, I. A. "K harakteristike agrarnih problemm partii II Internatsionala." *Agrarnye problem*, No. 3 (1928), 5–25.
Shanin, Teodor. *The Awkward Class: Political Sociology of Peasantry in a Developing Society: Russia 1910–1925*. Oxford: Clarendon Press, 1972.
Shanin, Teodor. *Defining Peasants: Essays Concerning Rural Societies, Expolary Economics, and Learning from Them in the Contemporary World*. Oxford: Basil Blackwell, 1990.

Shanin, Theodor. *Peasants and Peasant Societies: Selected Readings*. Harmondsworth: Penguin, 1971.

Shest *let borbi za krestianstvo. Obzor mezhdunarodnogo krestianskogo dvizhenia. Sb. Statei*. Moskva: Mezhdunarodnyi agrarnyi institut, 1935.

Shest *let mirovogo agrarnogo krizisa v tsifrah i diagramah*. Moskva, 1935.

Simonds, F. H. *Histoire de l'Europe d'apres guerre. De Versailles au lendemain de Locarno*. Paris: Payot, 1929.

Smilov, Daniel, and Jurij Toplak. *Political Finance and Corruption in Eastern Europe: The Transition Period*. Aldershot: Ashgate, 2007.

Smith, Anthony. *National Identity*. Reno: University of Nevada Press, 1991.

Snyder, Louis. *The Meaning of Nationalism*. New Brunswick, NJ: Rutgers University Press, 1954.

Snyder, Timothy. *Bloodlands: Europe between Hitler and Stalin*. New York: Basic Books, 2010.

Snyder, Timothy. *The Reconstruction of Nations: Poland, Ukraine, Lithuania, Belarus 1569-1999*. New Haven, CT: Yale University Press, 2003.

Sorokin, P. A. Ideologiia *na agrarizma*. Sofia, 1924.

Šouša, Jiří. "K vývoji českého zemědělství na rozhraní 19. století. Česká zemědělská rada 1891-1914." In *Acta Universitatis Carolinae Philosophica et Historica*. Vol. XCVII. Prague: Universita Karlova, 1986.

Šouša, Jiří. "Zemědělská správa a některá východiska vzestupu agrární strany (1914 -1918)." In *Acta Universitatis Carolinae Philosophica et Historica*. Vol. IV-V. Z pomocných věd historických VI-VII. Prague: Universita Karlova, 1987, 81-119.

Šouša, Jiří, Daniel F. Miller, and Mary Hrabik Samal, eds. *K úloze u významu agrárního hnutí v českých a českoslobenských dějinách*. Praha: Univerzita Karlova, 2001.

Spulber, N. "Changes in the Economic Structures of the Balkans, 1860-1960." In Charles Jelavich and Barbara Jelavich, eds. *The Balkans in Transition: Essays on the Development of Balkan Life and Politics since the Eighteenth Century*. Berkeley: University of California Press, 1963, 346-75.

*Šta je seljačka internacionala*. Beograd: Štamparija Save Radenkoviča, s.d.

Stamboliiski, Aleksandur. *Printsipite na bulgarskiiat zemledelski suiuz (pisano v zatvora)*. Sofia: Pechatnitsa na Z. D. Vidolov, 1919.

Stamboliiski, Aleksandur. *Pulno subranie na uchineniiata mu*. Pod redaktsiata na Kuniu Kozhuharov. Sofia: Fondatsia Aleksandur Stamboliiski, 1947.

Stamboliiski, Aleksandur. *Izbrani proizvedeniia*. Edited by Nencho Dimov, Iordan Zarchev, and Bogomil Volov. Sofia: Izdatelstvo na BZNS, 1979.

Stanković, Đ. *Nikola Pašić i Hrvati (1918-1926)*. Beograd: BIGZ, 1995.

Stavrianos, L. S. *The Balkans since 1453*. New York: New York University Press, 1958.

Stavrianos, L. S. *Balkan Federation: A History of the Movement toward Balkan Unity in Modern Times*. Hamden, CT: Archon Books, 1964.

Stavrianos, L. S. "The Balkan Federation Movement: A Neglected Aspect." *The American Historical Review* 48/1 (October 1942), 30-51.

Stedman Jones, Gareth. *Languages of Class: Studies in English Working Class History, 1832–1982.* Cambridge: Cambridge University Press, 1983.

Stefanovich, M. "Sumarni pregled agrarne reforme ma Balkanu." *La Federation Balkanique*, No. 98. 11. IX (1928).

Stoikov, Atanas. *Bulgarskata karikatura: kratuk ocherk za neiniia put i za tvorchestvoto na Aleksandur Bozhinov, Iliia Beshkov, Aleksandur Zhendov, Boris Angelushev i Stoian Venev.* Sofia: Bulgarski Hudozhnik, 1970.

Sugar, Peter, ed. *Eastern European Nationalism in the 20th Century.* Washington DC: American University Press, 1995.

Sugar, Peter, and Ivo Lederer, eds. *Nationalism in Eastern Europe.* Seattle: University of Washington Press, 1969

Sundhaussen, Holm. "Der Wandel in der osteuropäischen Agrarverfassung während der Frühen Neuzeit." *Südost-Forschungen* 2 (1990), 15–56.

Sundhaussen, Holm. "Die verpaßte Agrarrevolution. Aspekte der Entwicklungsblockade in den Balkanländern vor 1945." In Roland Schönfeld, ed. *Industrialisierung und gesellschaftlicher Wandel in Südosteuropa.* München: Verlag Otto Sagner, 1989, 45–60.

Sundhaussen, Holm. "Die Transformation des Dorfes und der Landwirtschaft im Balkanraum vom 19. Jahrhundert bis zum Zweiten Weltkrieg." *Südosteuropa Mitteilungen* 4 (1996), 319–35.

Sundhaussen, Holm. "Vom Vor- zum Frühkapitalismus. Die Transformation des Dorfes und der Landwirtschaft im Balkanraum vom 19. Jahrhundert bis zum zweiten Weltkrieg." In Frank-Dieter Grimm and Klaus Roth, eds. *Das Dorf in Südosteuropa zwischen Tradition und Umbruch.* München: Südosteuropa-Gesellschaft, 1997, 29–48.

Tasić, Nikola, and Miodrag Maticki, eds. *Svetozar Marković I Ljuben Karavelov u kontekstu slovenske književnosti i kulture.* Beograd: Balkanoloshki institute SANU, 1992.

Teslaru-Born, Alina. *Ideen und Projekte zur Föderalisierung des Habsburgischen Reiches mit besonderer Berücksichtiging Siebenbürgens 1848–1918.* Frankfurt-am-Main: Johann-Wolfgang-Goethe-Universität zu Frankfurt am Main, 2005.

Tibal, A. *La crise des Etats Agricoles Europeens et l'action internationale.* Paris: Publications de la Conciliation Internationale, 1931.

Timov, S. *Ekonomika Vostochnoi Evropi. Agrarizatsiia ili industrializatsiia.* Moskva: Gosudarstvennoe sotsialno-ekonomicheskoe Izdatelstvo, 1931.

Todorov, Kosta. *Balkan Firebrand: The Autobiography of a Rebel, Soldier and Statesman.* Chicago, IL: Ziff-Davis, 1943.

Todorov, Varban. *Greek Federalism During the Nineteenth Century (Ideas and Projects).* Boulder, CO: East European Monographs, N. CDVIII, 1994.

Todorova, Maria. "Balkanism and Postcolonialism, or On the Beauty of the Airplane View." In Costica Bradatan and Sergei Oushakine, eds. *In Marx's Shadow: Knowledge, Power, and Intellectuals in Eastern Europe and Russia.* Lanham, MD: Lexington Books, 2010, 175–96.

Todorova, Maria. "The Course and Discourses of Bulgarian Nationalism." In Peter Sugar, ed. *Eastern European Nationalism in the Twentieth Century.* Washington DC: American University Press, 1995, 55-102.

Todorović, Desanka. *Jugoslavija i Balkanske države 1918-1923.* Beograd: Narodna knjiga. Institut za savremenu istoriju, 1979.

Tomasevich, Jozo. *Peasants, Politics and Economic Change in Yugoslavia.* Stanford, CA; London: Stanford University Press, 1955.

Tomasevich, Jozo. *War and Revolution in Yugoslavia, 1941-1945: Occupation and Collaboration.* Stanford, CA: Stanford University Press, 2001.

Tomitch, Z. *La formation de l'etat yugoslave.* Paris, 1927.

Trouton, Ruth. *Peasant Renaissance in Yugoslavia 1900-1950: A Study of the Development of Yugoslav Peasant Society as Affected by Education.* London: Routledge & Kegan Paul, 1952.

Tseovaridou, Theano. "La structure agricole de la Grece et de la Bulgarie a la fin du XIX et au debut du XX s." *Balkan Studies* 25/2 (Thessaloniki, 1984).

Tuleshkov, Ivan. *Deitsi na Yuzhnoslaviansko edinstvo.* Sofia: Zemia i kultura, 1948.

Tushunov, A. V. *Voprosy agrarnoi teorii.* Moskva: "Mysl," 1976.

Uhlíř, Dušan. *Republikánská strana zemědělského a malorolnického lidu 1918-1938: Charakteristika agrárního hnutí v Československu.* Práce oddělení novějších československých dějin. Vol. 1, No. 2. Ústav československých a světových dějin ČSAV, 1988.

Varga, I. M. "Nashe zadrugarstvo." *Jubilarni zbornik zhivota i rada SHS* 1.XII.1918-1.XII.1928 (Beograd, 1928).

Veber, France. *Idejni temelji slovanskega agrarizmu.* Ljubljana: Kmetijska tiskovna zadruga, 1927.

Velev, Aleksandur. *Glavni reformi na zemedelskoto pravitelstvo.* Sofia: Nauka i izkustvo, 1977.

Velev, Aleksandur. *Prosvetna i kulturna politika na pravitelstvoto na Aleksandur Stamboliiski.* Sofia: Izdatelstvo na BZNS, 1980.

Verdery, Katherine. *National Ideology under Socialism: Identity and Cultural Politics in Ceauşescu's Romania.* Berkeley: University of California Press, 1991.

Verdery, Katherine. *Transylvanian Villagers: Three Centuries of Political, Economic, and Ethnic Change.* Berkeley: University of California Press, 1983.

Verdery, Katherine. *The Vanishing Hectare: Property and Value in Postsocialist Transylvania,* Ithaca, NY; London: Cornell University Press, 2003.

Verdery, Katherine, and Sharad Chari. "Thinking between the Posts: Postcolonialism, Postsocialism, and Ethnography after the Cold War." *Comparative Studies in Society and History* 51/1 (2009), 6-34.

Vilcu, Ş. "Aspects de la réforme agraire en Yugoslavie." *Revue des etudes sud-est europeennes* 12/3 (Bucarest, 1984), 253-58.

Vinaver, Vuk. *Jugoslavija-Frantsuska izmedju dva rata.* Beogread: Institut savremene istorije, 1986.

Viola, Lynne, V. P. Danailov, N. A. Ivnitskii, and Denis Kozlov. *The War against the Peasantry 1927-1930: The Tragedy of the Soviet Countryside*. New Haven, CT: Yale University Press, 2005.

Vucinich, Wayne, and Jozo Tomasevich. *Contemporary Yugoslavia: Twenty Years of Socialist Experiment*. Berkeley: University of California Press, 1969.

Vučo, Nikola. "'Agrarni blok' podunavskih zemlia za vreme ekonomske krize 1929-1933." In Vasa Čubrilović, ed. *Svetska ekonomska kriza 1929-34 godine i nijen odraz u zemiama Jugoistochne Evrope*. Beograd: Srpska Akademija nauka i umetnosti, 1976, 29-53.

Vučo, Nikola. *Agrarna kriza u Jugslaviji. 1930-1934*. Beograd: Prosveta, 1968.

Vučo, Nikola. *Poljoprivreda Jugoslavije (1918-1941)*. Beograd: Izdavačko Preduzeče "Rad," 1958.

Weber, Eugen. *Peasants into Frenchmen: The Modernization of Rural France 1870-1914*. Stanford, CA: Stanford University Press, 1976.

West, Harry G., and Parvathi Raman. *Enduring Socialism: Explorations of Revolution & Transformation, Restoration & Continuation*. New York: Berghahn Books, 2009.

Winter, Jay. *The Legacy of the Great War: Ninety Years On*. Columbia: University of Missouri Press, 2009.

Winter, Jay. *Remembering War: The Great War between History and Memory in the 20th Century*. New Haven, CT: Yale University Press, 2006.

Winter, Jay. *Sites of Memory, Sites of Mourning: The Great War in European Cultural History*. Cambridge: Cambridge University Press, 1998.

Winter, Jay, and Blaine Baggett. *The Great War and the Shaping of the 20th Century*. New York: Penguin, 1996.

Wladigeroff, Theodor. "Agrarverfassung und Agrarprobleme in Bulgarien." *Berichte über Landwirtschaft* 13/4 (1930), 650-718.

Wolf, Erik. *Peasants*. Englewood Cliffs, NJ: Prentice-Hall, 1966.

Yovanovic, Dr. *Les effets economiques et sociaux de la guerre en Serbie*. Paris: Les Presses universitaires de France, 1930.

Yuval-Davis, Nira. *Gender and Nation*. London: Sage, 1997.

Zagoroff, S. D. "Rise and Decline of peasant freedom in the Danubian Countries." *Weltwirtschaftliches Archiv* 69/2 (1952), 274-310.

Zebitch, Milorade. *La Serbie agricole et sa democratie*. Paris, 1917.

Zering, M. *Agrarnaiia revoliutsiia v Evrope*. Moskva, 1926.

Zering, M. *Agrarnie krizisi*. Moskva, 1927.

Zharnovskii, Ia. "Problema avtoritarnih i diktatorskih rezhimov v Tsentralnoi i Vostochnoi Evrope v periode mezhdu dvumia mirovymi voinami." *Etudes Balkaniques*, No. 2 (Sofia, 1973).

Zotschew, Todor. "Wechselbeziehungen von Sozialstruktur und Außenhandel in den südosteuropäischen Ländern." In Franz Ronneberger and Gerhard Teich, eds. *Von der Agrar—zur Industriegesellschaft. Sozialer Wandel auf dem Lande in Südosteuropa*. Darmstadt: Hoppenstedt, 1969, 1-34.

# Index

Abrashev, Petur  78
Agrarianism  7–19, 33, 39, 47, 55, 61, 93, 97, 101, 103–7, 112–21, 151, 153, 156, 159, 161, 164, 167–73
   definition  10, 17, 103–7, 113, 116, 120
Alexander, King of Serbia  145, 147, 148
Alexandrov, Todor  95
Anderson, Benedict  68, 86, 89, 90
Andreichin, Liubomir  82
Aprilov, Vasil  72, 74
Atanasov, Nedialko  58, 96, 97
Ausgleich  52
Austria  15, 49, 52, 55, 125
Austria-Hungary  4, 41, 55, 63
Austromarxists  52
Avramović, Mihailo  34, 45, 179 n.38

Balabanov, Alexander  81
Balibar, Étiene  68
Balkan Alliance systems  52, 54
Balkan Entente  54
Balkan federation  48, 52–4, 59, 146
Balkans  15, 33, 41–4, 47–8, 52, 57, 90, 105, 108, 121, 153, 154
Balkan Wars  12, 24–6, 37, 62, 82
Banac, Ivo  61, 183 n.1
Belgrade  13, 50, 53, 59, 73, 81, 96, 97, 122, 142–7, 150, 192 n.4
Bell, John  7, 25, 49, 63, 107, 111, 177 n.14
Beneš, Edvard  41, 50, 164, 199 n.33
Bengal  89
Beran, Rudolf  164–7, 200 n.39
Berend, Ivan  154
Berlin  4, 60, 186 n.39
Beron, Petur  72
Beshkov, Ilia  3, 175 n.4
Biondich, Mark  7, 35, 59, 122, 145
Blagoev, Dimitur  30, 31
Bobchev, Stefan  78
Bogosavljević, Adam  33

Boiadzhiev, Grigor  132, 133
Boiadzhiev, Ivan  124
Boiadzhieva, Nadezhda  131
Boll, Michael  158, 198 n.12
Bolsheviks  41–3, 46, 59, 96, 118, 149, 192 n.76
Boris III, King of Bulgaria  3, 24, 30, 63, 127, 128, 130, 193 n.28
Bosnia  12, 33, 34, 114
Botev, Hristo  73
Bozhinov, Aleksandur  3–4, 175 n.4
Brătianu, Ion  47
Brno  98, 99
Brubaker, Rogers  68
Bucharest  24, 42, 50, 53, 73, 185 n.32
Budapest  5, 50
Bukharin, Nikolai  119
Bulgaria  1–18, 21, 24–34, 38–9, 41–56, 58, 60, 62–4, 66–7, 92, 95, 97–102, 104, 106, 109–12, 114–18, 120, 122, 141, 150, 151, 163, 155–60, 162–7, 169–72
   orthographic reform  71–86
Bulgarian Academy of Sciences (BAS)  3, 83, 112
Bulgarian Agrarian National Union (BANU)  3, 6, 8, 12, 15–19, 21–2, 24–33, 37–8, 45–6, 47–9, 55, 57–8, 60, 64, 66, 78, 82–4, 95–101, 106–11, 114–16, 119, 122, 141, 156–60, 162, 166, 170–1, 187 n.62, 190 n.39, 198 n.14
   "Aleksandur Stamboliiski"  19, 158, 159, 162
   BANU-"Nikola Petkov"  158
   BANU-United  158
   corruption trial  123–34
   *druzhbi*  31, 99, 101, 122–4, 126–7, 188 n.6, 193 n.13
   labor service  64, 107–9, 190 n.39
   Orange Guard  32
   "Pladne"  97, 156, 157, 160

reforms 18, 24, 63–4, 81–5, 107, 111–12, 101, 118, 122, 159, 191 n.51
Representation of BANU Abroad 17, 58, 95–9, 116, 126–7, 166, 171, 188 n.9
Union of the Bulgarian United (*zdruzheni*) Agrarians Abroad 17, 98–101
"Vrabcha 1" 65, 97, 156
Bulgarian Communist Party (BCP) 12, 30–2, 47, 53–4, 57–8, 84–6, 100, 104, 112, 141, 150, 156–8, 197 n.96
Bulgarian Social Democratic Workers' Party 30
*Bulgarska komunisticheska partiia*, see Bulgarian Communist Party
*Bulgarska rabotnicheska sotsialdemokraticheska partiia*, see Bulgarian Social Democratic Workers' Party
*Bulgarski Zemedelski Naroden Suiuz (BZNS)*, see Bulgarian Agrarian National Union (BANU)
*Bulletin Mezinárodního Agrárního Bureau (Bulletin of the MAB)* 13, 17, 44, 45, 95, 101, 112–19, 165

Central Europe, see Eastern Europe
*Československá strana agrární*, see Czech-Slavic Agrarian Party
*Československá strana zemědělská*, see Czechoslovak Agrarian Party
Četniks 160
Chakrabarty, Dipesh 87, 89
Chari, Sharad 89–90
Chatterjee, Partha 62, 86–92, 153
Chayanov, Alexander 6, 175 n.7
Chibber, Vivek 87–8, 90
Church Slavonic 71–2, 74, 81
Coandă, Constantin 47
Cold War 14, 68, 89–90, 137
Comintern 14, 53, 56–9, 119, 143, 145–7, 182 n.49
cooperative movement 17, 34, 46, 57, 64, 106, 109–11, 114–15, 117–20, 159, 163, 173, 179 n.37, 190 n.48
corruption 7, 11, 18, 49, 120–7, 131–42, 171–2, 195 n.56
Croatia 7–9
Croat People's Peasant Party (CPPP) 12, 34, 35, 38, 172

Croat Republican Peasant Party (CRPP) 12–13, 18, 22, 35, 39, 59, 122, 143–150
*Crvena Pravda* 151
Czechoslovak Agrarian Party 163
Czechoslovakia 6–9, 11–13, 15, 17, 19, 21–3, 30, 32, 35, 39, 41, 45, 46, 50–2, 55–6, 58–60, 92, 95–6, 101, 104, 106, 113–17, 119, 153, 156, 163–5, 167, 169, 171, 199 n.28
Czechoslovak Republican Party 15, 104
Czech Republic 7, 9, 13, 19
Czech-Slavic Agrarian Party 37

Danailov, Georgi 78
Danev, Stoian 25, 78
Danubian Federation 52
Daskalov, Raiko 29–30, 95, 108, 112, 126
Davidović, Ljubomir 142, 144, 145, 196
*Demokratska Stranka* 144
Dewey, John 111
Dimitrov, Aleksandur 25
Dimitrov, Georgi 158
Dimitrov, G. M. 103, 156–8, 160
Dimou, Augusta 33
Dobrudzha 24, 28
Đokić, Radosav 59
Dorev, Pancho 97
Dragiev, Dimitur 25, 28, 30
Drašković, Milorad 151
Drinov, Marin 76, 80, 83–5
*Duma na bulgarskite emigranti* 73
*Dunavski lebed* 73

Eastern Europe 9, 14, 17, 21, 41, 52, 62, 85, 92, 104, 154, 158, 161–6, 171
Eastern Thrace 24
East-European studies 10, 16, 19, 66–70, 89–90, 105–6, 138, 148, 153, 170
Emelianoff, Ivan 114–15
England, see Great Britain
Estonia 56

Fadenheht, Iosif 78
Fellner, Fritz 23
Ferdinand, King of Bulgaria 3, 24–6, 28, 30, 47, 178 n.10
First World War 1, 2, 6, 11, 13–14, 19, 21–5, 31–3, 37–9, 47, 51, 53, 55–6, 76, 79–80, 82, 86, 104, 113, 115, 127, 131–2, 134, 169

Fleurant-Agricola, Gabriel  101–2, 118
France  41, 49, 73, 91, 101, 118, 144, 145, 157, 160
Frankenberger, Otokar  119
Franz-Ferdinand  4

Ganev, Venelin  78
Gellner, Ernest  68, 105
Genadiev, Nikola  63
Genoa Conference  49, 181 n.31
Genovski, Mihail  45, 180 n.13, 192 n.12, 193 n.32
Georgiev, Kimon  45
Germany  27, 52, 54, 111, 114, 141, 164, 165
Gerov, Naiden  75
Gichev, Dimitur  97
Great Britain  50, 111, 144, 145, 157, 164
Grebenarov, Alexander  128, 194 n.34
Greece  47, 50, 54, 55, 158
Greenfeld, Leah  68, 70–1
Green International  8, 9, 13–15, 17, 19, 41–7, 50–1, 55–6, 59, 60, 102, 105, 112–13, 115–16, 119, 160, 165–7, 169–70, 182 n.43
Gruev, Ioakim  75

Hácha, Emil  164
Harlakov, Ivan  127–9
Hayes, Carleton  67
Hodža, Milan  7, 9, 15, 44, 55, 118, 163, 166
Hroch, Miroslav  73
*Hrvatska Pučka Seljačka Stranka*, see Croat People's Peasant Party (CPPP)
*Hrvatska Republikanska Seljačka Stranka*, see Croat Republican Peasant Party (CRPP)
Hungary  23, 41, 42, 52, 55, 164

Iliev, Dimitur  46
Independent Agrarian Party (Slovenia)  34
International Agrarian Bureau (*MAB*), see Green International
International Institute of Agriculture, Rome  60
International Peasant Council, see Krestintern
International Peasant Union  160–1, 166–7

Ionesco, Ghita  105
Ivanchev, Todor  76, 80, 83–5

Jasenovac  160
Jelavich, Barbara  153
Jovanović, Dragoljub  160, 162
Jovanović, Jovan  44, 45

Karadžić, Vuk  72
Karavelov, Liuben  75, 76, 185 n.32
Kepka, Josef  165–6
Kharkiv  75, 115
Kingdom of Serbs, Croats, and Slovenes (Triunine Kingdom), see Yugoslavia
Kligman, Gail  155
Klindera, Ferdinand  101, 114, 115
Kohn, Hans  67–9
Koraïs, Adamantios  72
Kramář, Karel  36, 51
*Krestintern*  13–15, 18, 47, 56–60, 116, 119, 122, 144–9, 170, 172, 182 n.49
Kun, Bela  42
Kushev, Dimitur  130

*La fédération balkanique*  53
Latin America  57, 91
Latvia  56
Liapchev, Andrei  96–8
Little Entente  41–3, 46, 50, 52, 180 n.1
Livezeanu, Irina  69
Ljubljana  116
London  50, 59, 143–4, 149
Lulchev, Georgi  132

Macedonia  24, 29, 53, 54, 82, 95, 123, 129, 192 n.4
Maček, Vladko  7, 19, 59, 142–4, 147, 160–1
Maier, Charles  23
Malinov, Alexander  28, 30–1
Marković, Svetozar  33
Masaryk, Tomáš  36, 51
Matković, Hrvoje  148
Mečiar, Vladimir  163
Mečíř, Karel  44–5, 50, 101–2, 112, 117
*Mezinarodni Agrarni Bureau*, see International Agrarian Bureau (*MAB*)
Midhat pasha  118

# Index

Mihailovski, Nikola 75
Mikołajczyk, Stanisław 160
Miletich, Liubomir 76–8, 84
Mishaikov, Dimitur 78
Mollov, Vladimir 78
Momchev, Stancho 31
Momchilov, Ivan 75–6
Moscow 11, 13, 18, 32, 47, 54, 57–9, 122, 144–5, 149–50, 170, 172

Nairn, Tom 68
*Narod* 123
*Narodna Radikalna Stranka*, see National Radical Party (NRS)
*Narodno Subranie* 31, 130
nationalism 7, 11, 12, 16, 17, 24, 39, 52, 60–71, 76, 80, 82, 84–6, 88–9, 91–3, 118, 153, 169–71
National Radical Party (NRS) 12, 33, 39, 145
Neofit Rilski 72
New Economic Policy (NEP) 56
*Nezavisimost* 83, 185 n.32
*Nová agrárna strana* 163
*Nova Bulgariia* 73

Obbov, Aleksandur 49, 58, 95–8, 114–15, 118
*Obznana* 122, 143, 150–1
October Revolution 23
Omarchevski, Stoian 24, 76–85, 111, 186 n.43
Ottoman Empire 22, 52, 71, 185 n.22, 186 n.39

Paisii Hilendarski 71
Panizza, Francisco 104
Pašić, Nikola 12, 33, 47, 48, 117, 144, 147–9, 196 n.78
Paxton, Robert 102
*Pětka* 36, 38, 180 n.43
Petkov, Dimitur 157
Petkov, Milo 132
Petkov, Nikola 25, 27, 157–8, 178 n.10, 198 n.13
Petkov, Petko 157
Petrova, Dimitrina 83, 134
*Piast* 115, 160
Poincaré, Henri 49

Poland 55–6, 69, 102, 114–16, 125, 141, 160, 167
Popović, Mihailo 34
Pop-Petrov, Peicho 98–9, 103
Prague 42, 44–5, 50–1, 95, 97, 101, 114–16, 127, 165
*Priaporets* 83, 122
Protić, Stojan 33

Radić, Pavle 147
Radić, Stjepan 7, 8, 12–13, 18, 35, 38, 54, 59, 117, 121–2, 142–9, 156, 172
*Radikal* 83
Radomir Rebellion 29–30, 95, 127, 179 n.26
Radoslavov, Vasil 25, 27, 178 n.17
Rakovski, Georgi Sava 73, 186 n.43
Renč, Václav 165–6
Republican Party of Agriculturalists and Small Farmers, see Slovak National and Peasant Party
Republican Party of the Czechoslovak Countryside (RSČV) 13, 22, 35, 39
*Republikánská strana československého venkova*, see Republican Party of the Czechoslovak Countryside (RSČV)
*Republikánská strana zemědělského a malorolnického lidu*, see Republican Party of Agriculturalists and Small Farmers
Reviakina, Luiza 145, 182 n.49
Rističević, Marijan 163
Romania 26, 41, 47, 50, 52, 55, 69, 75, 102, 114–17, 138–9, 145, 156, 161, 166
Romanski, Stoian 83
Russia 9, 26, 28, 42–3, 54, 56–7, 74, 76, 91–2, 102, 110, 113–17, 138–9, 145, 161, 166

Sakuzov, Ianko 31, 52
*Samostalna Kmetijska Stranka*, see Independent Agrarian Party (Slovenia)
*Savez Težaka u Bosni i Hercegovini* 12, 34

*Savez Zemljoradnika (SZ)* 9, 11–13, 22–3, 32–4, 37, 39, 44, 62, 115, 117, 162
Scott, James C. 5, 139
Second World War 19, 45, 67, 86, 92, 97, 104, 112, 135, 141, 151, 154–6, 163–8, 170, 172
*Seljački Savez* 12, 34
*Selo* 34
September Uprising 30, 58, 124, 151, 157, 179
Serbia 7, 9, 11–13, 19, 22, 27, 33, 38–9, 47–8, 50, 59, 63, 72, 76, 81, 148–9, 160, 162–3, 169
Serbian Agrarian Party 59
Shanin, Teodor 5
Sharenkov, Andrei 31
*Skupština* 33–4, 144–5, 147, 150
Slovakia 7, 9, 19, 39, 163–4, 167
Slovak National and Peasant Party 13
*Slovenská národná a rolnícka strana*, see Slovak National and Peasant Party
Slavo-Bulgarian school 72
Snyder, Louis 67
Snyder, Timothy 69, 188 n.79
Sofia 4–5, 27–31, 44, 50, 77, 79, 81–2, 95, 123, 125–8, 130–1, 133, 157, 166
Soldiers' Uprising, *see* Radomir Rebellion
Sorokin, Pitirim 114
Southeastern Europe, *see* Balkans
Soviet Union 42, 104, 118, 143, 149, 155, 157, 160–1, 166
Stamboliiski, Alexander 1–10, 15, 18, 24–32, 37–8, 43–56, 62–4, 77, 79, 81, 84, 86, 104, 107, 109, 111, 116, 121–35, 140–2, 155–8, 171, 179 n.23, 180 n.13, 191 n.51, 192 n.4, 194 n.45
Stamboliiska, Milena 131
Stambolov, Stefan 75–6
Stoianov, Hristo 58
Stoianov, Petko 78
subaltern studies 11, 16–17, 60, 62, 67–8, 86–92, 153, 171
Sugar, Peter 69
Syrový, Jan 164
*Svaz slovanské agrární mládeže*, see Union of Slav Agrarian Youth

Švehla, Antonín 36, 38, 44–5, 51, 55, 114–15, 117–18, 156
Sweden 114

Teodorov, Teodor 31
Teodorov-Balan, Alexander 76, 80–1
*Težački Savez* 12, 34
Todorov, Kosta 27, 30, 49, 58, 96, 97, 166
Tomov, Kosta 126
Treaty of Neuilly 24, 42, 50, 64, 82, 108
Treaty of Niš 53, 55
Treaty of Sèvres 42
Trifonov, Stancho 98
Triunine Kingdom, *see* Yugoslavia
Tsankov, Alexander 18, 45, 58, 78, 83–4, 96, 98, 115, 122, 124, 129–30, 133–4, 151
Tserkovksi, Tsanko 37
Tsitsonkov, Iordan 95
Tsonev, Benio 76, 80–1
Tusar, Vlastimil 36

Union of Agriculturalists (UA), *see Savez Zemljoradnika*
Union of Democratic Forces 158
Union of Slav Agrarian Youth 116
United States 52, 111, 114, 154, 157–8, 160
Ustaša 160

Velestinlis, Rhigas 52
Velev, Alexander 81
Venizelos, Eleftherios 47–8
Verdery, Katherine 89–90, 139, 155
Versailles system 15, 46–7, 55–6, 170
Vidovdan constitution 38, 54, 144, 147, 149
Viškovský, Karel 115
Vlajinac, Milan 34
Vucinich, Wayne 143
Vulkov, Georgi 119
Vulkov, Ivan 128–9, 167
*Vuzroditelen Protses* 141

Warsaw 42, 50, 116–17
Western Thrace 24
Wiles, Peter 105
Wolf, Eric 5

Wrangel, Pyotr   43
*Wyzwolenie*   115, 117

Yugoslavia   6, 8–9, 11–12, 18, 32, 39, 45,
    50–5, 58–9, 61–2, 92, 106, 114–18,
    146, 149, 151, 153, 156, 162, 167,
    169, 172
  Triunine Kingdom   21, 38–9, 41, 50,
    82, 95, 142–5, 149–50

*Zeleni Kadar*   34
*Zemledelska Zashtita*   126
*Zemledelsko Zname*   24–5, 27–8, 43, 78,
    83, 95–6, 122, 126–8
*Zemljoradnička Stranka*, *see* Serbian
    Agrarian Party
Zhivkov, Todor   141
*Zname*   73
Zveno   97, 130, 157

www.ingramcontent.com/pod-product-compliance
Lightning Source LLC
Chambersburg PA
CBHW052036300426
44117CB00012B/1849